Proceedings of a Meeting of the EC Experts' Group / Dublin 12-14 June 1985

WEED CONTROL ON VINE AND SOFT FRUITS

Edited by

R.CAVALLORO
Commission of the European Communities, Joint Research Centre, Ispra

D.W.ROBINSON
Agricultural Institute, Kinsealy Research Centre, Dublin

Published for the Commission of the European Communities by

A.A.BALKEMA / ROTTERDAM / BOSTON / 1987

*The texts of the various papers in this volume were set individually
by typists under the supervision of each of the authors concerned.*

Publication arrangements: *P.P.Rotondó,* Commission of the European Communities,
Directorate-General Telecommunications, Information Industries and Innovation, Luxembourg

EUR 10082

ISBN 90 6191 691 7

Published by A.A.Balkema, P.O.Box 1675, 3000 BR Rotterdam, Netherlands
Distributed in USA & Canada by A.A.Balkema Publishers, P.O.Box 230, Accord, MA 02018

ORGANIZING COMMITTEE

Cavalloro Raffaele, Meeting's responsible
Principal Scientific Officer CEC Programme "Integrated Plant Protection"

Robinson David, Meeting's chairman
Head Kinsealy Research Centre An Foras Taluntais

STRUCTURE OF THE MEETING

Introduction

Opening address by R. Cavalloro and D. Robinson

Sessions

Weed control in vines	Chairman: C.N. Giannapolitis
Weed problems in soft fruits	Chairman: D. Seipp
Weed control in strawberries	Chairman: G. Noye
Evaluation of herbicide tolerance	Chairman: H.M. Lowson
Integrated pest management	Chairman: D. Robinson

Conclusions

General discussion and closing address by Organizing Committee

ORGANIZING SECRETARIAT

Campion Gerard & Murray Pauline
Agricultural Institute, Kinsealy Research Centre, Dublin

Foreword

During the last forty years the science of weed control has grown from a position of relative insignificance to become one of the major technologies of modern agriculture. Until recently the problem of controlling weeds in vines and soft fruits was particularly difficult because the long-term nature of these crops allowed perennial weeds to become established over a period of years. The introduction of herbicides has considerably reduced for extent of this problem and has resulted in greatly increased yields.

Despite significant advances the problem of controlling weeds cheaply and effectively in vines and soft fruits is still far from solved. "New" weeds resulting from changing agricultural practices and interspecific selection by herbicides continue to appear. In addition, several important weed species have become resistant to herbicides to which they were previously susceptible. Moreover, where weeds have been well controlled, the resulting bare soil is more easily eroded in some areas.

In implementing a decision of the Council on the programme "Energy in Agriculture", the Programme Committee decided, as part of coordinated activities on Integrated Plant Protection, to consider weed problems emerging at CEC level. Accordingly, a meeting on "Weed control in vine and soft fruits" was organised in Dublin, June 12-14, 1985.

The problems already mentioned are common both to vineyards and soft fruit plantations and, as the same herbicides and strategies are used in both sectors of fruit growing, it was considered that an exchange of views between the two groups would be useful and timely. The purpose of the meeting was to review present knowledge on the use of herbicides in fruit plantations in order to identify suitable strategies for weed control and to develop systems of integrated methods of weed management.

The importance of these crops in the ten countries of the European Communities is shown by the fact that vineyards covered a total area of 2.5 m ha in 1983 and wine contributed 5.8% of final agricultural production. While grape production is restricted mainly to France, Italy, Germany, Greece and Belgium, strawberries (0.41 m ha) and other soft fruits (0.48 m ha) are grown in all ten countries. Eight EC countries (Belgium, Denmark, Germany, Great Britain, Greece, Ireland, Italy, Netherlands) and Switzerland, were represented at the meeting and delegates from all countries participated fully in the informative discussions both in the lecture room and in the field.

The papers given at the meeting are presented in full in these Proceedings and show the importance of weeds in fruit crops. In the appendix are the results of a questionnaire, that was circulated after the meeting, summarizing a list of the worst weeds in vine and soft fruits in each European Country. The papers also demonstrated clearly the common problems faced by fruit growers throughout the Community and the 'Conclusion and Recommendations' show the scope that exists for further work in this area.

R. Cavalloro, D.W. Robinson

Table of contents

Session 4. *Evaluation of herbicide tolerance*

Session 5. *Integrated pest management*

Introduction

Opening address

R.Cavalloro
CEC, Joint Research Centre, Ispra, Italy

Distinguished and dear Colleagues,

I have the pleasure to welcome you in the name of the Commission of the European Communities, the General Directorate Agriculture, and in my own name, and I would like to express my keen satisfaction for your highly qualified participation in this important EC-meeting.

This meeting follows a similar experts' meeting which dealt with the weed problem, organised similarly by the CEC and held in Braunschweig (FRG) in October 1980 on "The rationalization of herbicide use". Six European Countries were represented (Belgium, Denmark, F.R. Germany, Great Britain, Ireland, Netherlands), and the effects of herbicides on the environment were considered. In particular, some problems such as competition, aggressivity, interference with harvesting, population dynamics, the tolerance of crops to herbicides, as well as combining chemical with mechanical of biological methods to reduce herbicide use, and the cost/benefit analyses were thoroughly examined.

For different reasons, we must use chemicals with discretion. The manufacturers can contribute to this aim by tailoring herbicides with less side-effects on human beings and on the environment as well.

A Council decision on agricultural joint programmes was started at the ten European Countries level. It concerns research on Integrated and Biological Control, with the aim of a more sensible utilization of pesticides based on biological knowledge, and on an overall reduction in the use of chemical products. Since December 1983, by adopting joint research programmes and programmes for coordinating agricultural research, the Community grants financial aid for projects in the field of Integrated Plant Protection, developing effective methods of controlling pests, weeds and diseases of plant, involving natural control agents which are less harmful to man and to environment and which use less energy. Among various other subjects, weeds are of remarkable importance, relating to improving integrated control for different crops, where the frequent development of resistance requires a constant search for new active substances and the development of new strategies of intervention.

3

In this context our meeting considers the weed problem in vine and in soft fruits. It is centered on two objectives: 1) to review and increase our present knowledge in order to identify suitable actions for weed control and for the rationalization of herbicide use, and 2) to look for a basis for future systems for integrated methods of weed management.

The engagement of all the participants will contribute, I am sure, to explore these objectives in depth, and aid the progress of agriculture production while maintaining respect for human health and the environment.

I wish this meeting, at this prestigious St. Patrick's College in Dublin, full success, and I would like to thank warmly Dr. D.W. Robinson and his collaborators for their spirit of cooperation in organizing so faultlessy the EC-meeting "Weed control in vine and soft fruits".

Weed control in soft fruits in Ireland – Introduction and overview

D.W.Robinson

Kinsealy Research Centre, Agricultural Institute, Dublin, Ireland

Summary

In Ireland the mild moist climate is favourable for the growth of weeds throughout the year. Investigations on the use of herbicides in fruit crops began at Loughgall, Co. Armagh in 1955 and at Clonroche, Co. Wexford in 1962. Good control of weeds has been achieved in all crops by using flexible programmes according to the weed species present or anticipated. Studies on the effect of repeated annual applications of herbicides and the elimination of tillage showed no significant adverse effect on soil structure. However, considerable change has occurred in the weed flora in fruit plantations during the last 30 years. In the 1950s and early 1960s the most prevalent species included Poa annua, Senecio vulgaris, Stellaria media and Ranunculus repens. These species have been replaced to some extent in the 1980s by weeds such as Galium aparine, Viola arvensis, Epilobium ciliatum, Vicia sativa, Convolvulus arvensis and volunteer crops.

Work in the 1950s and 1960s gave results that were quickly taken up by fruit growers. However, some practices, e.g. the use of simazine on charcoal-dipped strawberry runners, have been superceded by more convenient treatments.

Because of significant changes in the composition of weed populations and differences in herbicide tolerance between fruit cultivars, there is a need for a wide range of herbicides to deal with specific problems. Some of the early introduced herbicides and practices may still have a role in certain situations as a result of changing circumstances.

I. Introduction

In the 1950s and 1960s weeds were recognised as one of the major problems affecting the production of soft fruits in Ireland. The mild, moist climate encourages weed growth throughout the year and, before the advent of herbicides, growers faced serious difficulties in wet periods and during the winter when soil tillage was impossible.

2. Climate

The dominant features of the Irish climate are:

(a) relatively high annual rainfall consisting of non intensive rainfall of frequent occurrence

(b) relatively low duration of bright sunshine

(c)   relatively low insolation

(d)   low evaporation

(e)   moderate summer temperatures and mild winters.

2.1 Rainfall

Most of the eastern half of the country receives between 800 and 1,000 mm/annum and on average rain falls on approximately 180 days per annum. The western part of the country is wetter with an average of 220 wet days and a rainfall ranging from 1,000 to 1,200 mm. Rainfall is much higher in some mountainous regions.

2.2 Sunshine

Bright sunshine averages about 30% of the total possible. It is lowest in the north west (about 28% of maximum) and increases gradually towards the south east (about 32% of maximum).

2.3 Insolation

Because of location (53-55 NL), total solar radiation amounts to about 350kj/cm^2 which is approximately half of that received at the equator. Total radiation does not differ much from that in western and north eastern Europe but there are local variations due to differences in cloud cover.

2.4 Evaporation

Because of the high and frequent rainfall and low sunshine, evaporation is low at about 400 mm/annum from a grass cover. Most of the evaporation occurs from April to September inclusive. The low evaporation has advantages in relation to availability of water for crops and disadvantages in relation to fungal diseases as well as hindering weed control.

2.5 Temperatures

Temperatures are moderate with summer maxima seldom exceeding 22°C; however, an average daily temperature of 13-15°C is maintained during the months of June to September which is suitable for the cultivation of a range of temperate fruit crops including strawberries, blackcurrants, raspberries, gooseberries and blueberries. The climate is marginal for vine cultivation and unsuitable for <u>Actinidia chinensis</u>. Winter temperatures are moderate with a January mean at Clonroche, Co. Wexford of approximately 6°C.

3.   Soft fruit growing in Ireland

In 1985 the total area of open field-grown strawberries in the Republic of Ireland is approximately 500ha. Slightly over half of this area is grown for processing and the remainder for the fresh market. Over 80% of the strawberries for processing are produced in Co. Wexford in the south east of Ireland. Most of the strawberries for the fresh market are grown in counties close to Dublin and Cork, the major centres of population. 'Pick-your-own' fruit plantations occur in many parts of the country but this system of marketing is not as popular as in the United Kingdom. About 2,000 tonnes of strawberries are exported each year mainly as pulp for processing. This trade is under threat from cheap imports of pulp and frozen fruit from Poland and other east European countries.

Blackcurrants, raspberries and gooseberries are also grown in Ireland for processing and the fresh market but these crops are less important than strawberries. Blueberries grow well in Ireland and have been produced successfully on an experimental basis since 1955 but there has been little commercial uptake. Although vineyards occur in several part of the country and good quality wine can be produced in warm seasons, wine production is not economic.

4. Research on weed control in soft fruits

Work on the use of herbicides for weed control in strawberries, raspberries, blackcurrants and gooseberries began at the Horticultural Centre, Loughgall, Northern Ireland in 1955. Parallel research at the Soft Fruit Research Station, Clonroche, Co. Wexford began in 1962. Since 1956 a wide range of herbicides has been tested in soft fruits and details of the results obtained are presented in the Annual Research Reports of both stations.

With the introduction of many new herbicides and, in particular, simazine and paraquat, rapid progress was made and, by the late 1950s, it was possible to obtain a relatively weed-free environment in bush and cane fruit plantations. This had not been possible previously using traditional cultural methods. As a high standard of weed control could now be obtained by herbicides alone, a number of experiments was conducted to investigate if soil cultivation was still necessary. Experiments were conducted on raspberries - 5 years (Robinson 1964a), strawberries - 4 years Robinson 1964b), blackcurrants - 5 years (Robinson and Allott 1968), gooseberries - 7 years (Allott et al 1971) and apples - 17 years (O'Kennedy and Robinson 1984).

No evidence was obtained that the absence of soil tillage caused a yield decrease in any fruit crop but not all crops showed a positive response. In general non-tillage systems gave yield improvements with blackcurrants, gooseberries and apples but not with raspberries and strawberries.

4.1 Soil structure

In all fruit plantations treated with herbicides and not cultivated, significant changes were recorded in many soil properties including pore volume, numbers and orientation of pores (Bulfin 1967), bulk density, infiltration capacity, soil strength (Bulfin and Gleeson 1967) and organic matter (Robinson 1975).

No serious adverse soil effect was observed in the absence of cultivation except that the increased bulk density and smoothness of the soil surface resulted in increased run-off and soil erosion on sloping sites. Runoff was greatly reduced and erosion virtually eliminated by the use of mulches of organic matter (Robinson 1964b). Although the use of a straw mulch has usually resulted in increased yields in Britain, lower yields of bush, cane and top fruits have often been recorded in Ireland where herbicides were used in conjunction with mulching. Reduced yields on mulched plots were due in some plantations to lower air temperatures and increased frost injury (Robinson 1966). The effect of a mulch in reducing soil temperatures during the growing season or in maintaining excessively moist conditions in the soil may also be responsible in some situation.

5. Changes in weed flora

Although no significant adverse effect on soil structure has been

7

recorded following the repeated annual application of herbicides, marked changes have occurred in the weed flora.

The changes that have occurred in the most prevalent weeds in herbicide-treated soft fruit crops in Co. Wexford during the last 25 years are shown in Table I.

Table I - Most common weeds in soft fruits in Co. Wexford in 1960 and 1985

1960	1985
Stellaria media	Galium aparine
Chenopodium album	Viola arvensis
Senecio vulgaris	Atriplex patula
Poa annua	Potentilla anserina
Agropyron repens	Epilobium ciliatum
Ranunculus repens	Vicia sativa
Cirsium arvense	Agropyron repens
Rumex spp	Convolvulus arvensis
Sinapis arvensis	Equisetum arvense
Chrysanthemum segetum	Trifolium repens
Capsella buras-pastoris	Sherardia arvense
Polygonum convolvulus	Hypericum humifusum
Galeopsis tetrahit	Aphanes arvensis
Lamium purpureum	Volunteer crops (cereals,
Polygonum aviculare	potatoes and blackcurrants

Source: N. Rath, Clonroche

Many of the species that were very common in 1960 e.g. _Poa annua, Senecio vulgaris, Stellaria media_ and _Ranunculus repens_ grow strongly during the autumn to spring period when control by cultivation is impracticable. These and many annual weeds in the 1960 list are no longer a problem having proved to be very susceptible to one or more of the commonly used herbicides in soft fruits, notably simazine, lenacil and phenmedipham in strawberries and to simazine and paraquat in bush and cane fruits. The only weed in the 1960 list that remains a problem in 1984 is _Agropyron repens_, but this species is not as serious as it was previously because of the use now of glyphosate as a pre-planting treatment.

The weeds that are now prevalent have been encouraged by a variety of factors. All have at least some degree of tolerance of the widely used herbicide simazine. Some of the "new" weeds such as _Galium aparine_ and _Viola arvensis_, germinate well in the autumn and may also have been encouraged by the swing to winter cereals which are frequently used in rotation with soft fruit.

Between 1955 and 1957 detailed records were also kept of the weed flora at the Horticultural Centre, Loughgall by listing in order of prevalence all weeds on sites used for experimental purposes. On an area of 8 acres, 63 species representing 5I genera were recorded. The most troublesome weeds were again recorded in a strawberry plantation at Loughgall between 1981 and 1983. The results of the surveys done in 1955/57 and 1981/83 are shown in Table 2.

Table 2 - Most common weeds in soft fruits, Loughgall, Northern Ireland - 1955-1957 and 1981-1983

(arranged in order of prevalence)

1955-1957	1981-1983*
Poa annua	Senecio vulgaris
Stellaria media	Stellaria media
Ranunculus repens	Trifolium repens
Senecio vulgaris	Poa annua
Holcus lanatus	Cardamine hirsuta
Agrostis stolonifera	Ranunculus repens
Cerastium vulgatum	Polygomum aviculare
Potentilla anserina	Capsella bursa-pastoris

*Recorded by B. Watters.

Senecio vulgaris, Stellaria media, Poa annua and Ranunculus repens were included among the most prevalent species both in 1955/1957 and 1981/1983. This land had been ploughed out of long-term grassland in 1981; consequently the change in weed species is not so marked as that experienced at Clonroche where the farm had been mainly arable during the last 25 years. It is noteworthy that brassica weeds Cardamine hirsuta and Capsella bursa-pastoris were present in 1981/83 although absent in 1955/57.

6. Need for changes in herbicide programmes

Because of significant changes in the composition of weed populations and marked difference in herbicide tolerance between cultivars, there is a need for a wide range of herbicides to deal with specific problems. Treatments that have been superceded by the development of more selective herbicides may still be useful in some situations as a result of changing circumstances. For example as newly planted strawberry runners were found to be susceptible to low doses of simazine applied shortly after planting, experiments were conducted to ascertain if the tolerance of strawberries to simazine could be increased by placing adsorbents in the vicinity of the root zone at planting time (Robinson 1965). No significant damage was recorded in several trials in which charcoal dipped runners were sprayed with simazine at 1.1 kg/ha a few days after planting. This treatment was adopted by a number of strawberry growers in Britain and Ireland during the 1960s. Within a few years, however, the use of activated charcoal plus simazine was superceded by the use of lenacil. This herbicide was safe on newly planted runners of many cultivars grown in the 1960s. Moreover, it could be used on non-charcoal dipped plants and so the dusty job of dipping runners was avoided.

The introduction of new lenacil-susceptible cultivars such as Ostara and the changing weed flora, resulting in a need to use sequences of different herbicides, suggest that activated charcoal could still be a useful root-dip treatment for strawberry runners. Part of the inconvenience of dipping runners in activated charcoal can be reduced by replacing the dry powder by a slurry made by mixing 1 kg of activated charcoal with 5 litres of water (Kratky et al 1970). This treatment is used routinely at present on strawberries planted in late summer and autumn by at least one large-scale strawberry grower in England (Goodwin 1985). The extra cost of using charcoal (material plus dipping) amounts to only UK £54/ha).

Examples of new uses for old herbicides can also be obtained from bush and cane fruits. In a series of experiments in the late 1950s and early 1960s, diuron gave satisfctory control of a number of annual weeds in soft fruits and was more effective than simazine or atrazine on some species. For example seedlings of Cardamine hirsuta (5cm wide) were killed by a dose of 0.5 kg/ha. However, diuron was shown to have a lower margin of safety than simazine or atrazine on blackcurrants, gooseberries and raspberries (Robinson 1962). For this reason and because of the wider spectrum of common annual weeds that were more readily controlled by simazine than by diuron e.g. Senecio vulgaris and Veronica persica, diuron was seldom used in soft fruit plantations in Ireland. As diuron is highly effective against weed species of increasing importance e.g. Cardamine hirsuta, Capsella bursa-pastoris and Epilobium spp, there are undoubtedly occasions when low doses of this herbicide could usefully be substituted for triazines in bush and cane fruits.

REFERENCES

1. ALLOTT, D.J., ROBINSON, D.W. and UPRICHARD, S.D. (1971). The response of gooseberries to non-tillage systems of management. Hort. Res. 1971, 11: 166-176.
2. BULFIN, M. (1967). A study of surface soil conditions under a non-cultivation management system. 11: Micromorphology and micro-morphometrical analysis. I. J. agric. Res. 6, 189-201.
3. BULFIN, M. and GLEESON, T. (1967). A study of surface soil conditions under a non-cultivation management system. 1: Physical and chemical properties. I. J. agric. Res. 6, 177-188.
4. GOODWIN, D.F. (1985). Personal communication.
5. KRATKY, B.A., COFFEY, D.L. and WARREN, G.F. (1970). Activated carbon root dips on transplanted strawberries. Weed Sci. 18: 577-580.
6. O'KENNEDY, N.D. and ROBINSON, D.W. (1984). Further results from trials with overall herbicides in apples at Ballygagin. Aspects of Applied Biology 8, 169-177.
7. ROBINSON, D.W. (1962). A progress report on trials with soil-applied herbicides in soft fruit crops. Proc. 6th British Weed Control Conf. 1962, 2: 609-617.
8. ROBINSON, D.W. (1964a). A comparison of chemical and cultural methods of weed control in raspberries. Proc. 7th British Weed Control Conf. 1964, I: 195-202.
9. ROBINSON, D.W. (1964b). Investigations on the use of herbicides for eliminating cultivation in soft fruits. Scientific Horticulture XVI, 1962-1963, 52-62.
10. ROBINSON, D.W. (1965). The use of adsorbents and simazine on newly planted strawberries. Weed Research, 5: I, 1965, 43-51.
11. ROBINSON, D.W. (1966). A comparison of chemical and cultural methods of weed control in gooseberries. Proc. 8th British Weed Control Conf. 1966, I: 91-102.
12. ROBINSON, D.W. (1975). Some long-term effects of non-cultivation methods of soil management on temperate fruit crops. Proc. XIXth International Horticultural Congress, Warsaw (1975), III: 79-81.
13. ROBINSON, D.W. and ALLOTT, D.J. (1968). A study of the spacing of blackcurrants under conditions of non-cultivation. Horticultural Research 1968, 8: 51-61.

Session 1
Weed control in vines

Chairman: C.N.Giannapolitis

New aspects of integrated methods of weed-management in Italian vineyards

A.Scienza
Istituto di Coltivazioni arboree, Università degli Studi, Milano, Italy

R.Miravalle
MOITAL, Milano, Italy

Summary

Due to new technical issues and adverse economic facts, Italian wine growers are paying more attention to weed control and soil management systems shifting from several tillage a year to less mechanical treatments, larger use of herbicides, grass-mulch or zero-tillage.

Trials carried out for several years have been followed to control the effetcs of different soil techniques on several parameters (yield, sugar, acidity, vigor).

An integrated system allowing spontaneous and selected weeds growing from harvest time to May followed by chemical weed control in Summer, may avoid weed competition and its negative effect on water and nutrition with less herbicides use, and without the negative impact on the soil structure caused by several mechanical treatments.

Introduction

In Italy 1 600 000 farmers manage 1 150 000 hectares of vineyards. This involves every climatic and pedological area, from sea-level to 1,300 metres above sea-level (Mount Etna, for example), with a thermal summation variation of between 1600°C and 5000°C and average rainfall ranging from 450/500 mm in the South to 1500 mm on the Eastern slopes of the Alps.

As far as the location of the growing area is concerned, 35% of vineyards is in the plains, 56% on the hills and 9% in the mountains. More than 400 cultivars are cultivated using about 32 root-stoks and more than 1000 different kinds of wine are produced, of which nearly 300 are ranked V.Q.P.R.D.

This leads to a wide and diversified range of planting systems (from 800 to 4000 vine-stocks per hectare), of farming and pruning techniques as well as, naturally, different aggressiveness by weeds and their specific control systems.

13

The most widely-used systems are the Espallier (440,000 ha.) especially in the North-West (Piemonte, Oltrepo' Pavese and Toscana), the Bush-tree (230,000 ha.) and the Tendone (220,000 ha.) in the South, the Sylvoz and its varietions (120,000 ha.), the Trellis (80,000 ha.) and the Bellussi (60,000 ha.).

Economic aspects

With a gross production value of more than 3,000 billion lira, vine-growing accounts for 9.8% of farming income. Wine production amounts to 70/80 million hectolitres to which must be added 1,400,000 tonnes of table-grapes.

Changes in life-style have led to a dramatic drop in per-capita wine consuption (from 110 litres in 1955 to 81 in 1983); competition due to "alternative" beverages (beer and soft-drinks) as well as certain difficulty in entering the foreign markets have brought about a surplus in table wines.

On the other hand, since production is so parcelled out, no ad hoc response to an adverse market trend is possible. Furthermore, energy dependence coupled with a few production factors have influenced the cost-proceeds ratio by adversely affecting the vine-growers' income.

There is, therefore, a general need for profitability to pick up again, which means a modernisation process in terms of mechanisation, phytosanitary defence and soil management. The problem of weed management in the vineyard is therefore tackled from an overall point of view, with particular attention being paid to agronomical aspects and by providing the vinegrowers with ad hoc economic solutions.

Soil Treatment

The different ecological areas have developed different soil-management techniques, but weed management is still carried out through several, repeated tillings (almost 80%). About 10% of vineyards are grassed over (all in the North) by resorting to an integrated system of mowing, tillings and weeding.

More than 20% of the wine-growing area is managed with herbicides which usually integrate with tillings and mowing (grass-mulch). To-day only 2% (a percentage which is rapidly increasing) of vineyards are managed in "no-tillage" by resorting to herbicides only.

The most widely-used mixed system is chemical control of weeds under the rows, whereas the inter-row is tilled or grassed over. Regional variations are marked, ranging from 5 to 50% of wine-growers.

14

The most important fact in the last few years - except for wider use of integrated system - is a qualitative one. The use of herbicides is less a stop-gap measure (in the past the use of desiccants was mostly intended to delay tillings than to replace them) and in situ, localized treatments agaist perennial species - however coupled with tillings - were confined to those vineyards which were most infested with weeds.

To-day the use of chemical herbicides is increasingly regarded as a soil-management method in line with a policy of "conservation-tillage" or "zero-tillage" so as to get economic, agronomical and technical gains.

UNSOLVED PROBLEMS

Traditional tillings

The need to develop some alternatives to traditional tillings stems from the inconveniences which this kind of tilling brings about, especially in clay soils which are fairly common in Italy.

Here, unsuitable tillings (with excess soil moisture) or use of unsuitable operating machines (milling machines or rotary hoes) cause the vineyards to suffer from so-called "compacting syndrome" as it is generally known (spring-time chlorosis, summer reddenings, vegetative and productive decay, sudden death of stocks); this can be ascribed to changes in the physical condition of the soil such as loss of porosity, compacting, anoxia, building up of harmful gases (ethylene, for example) in the soil, the tilling layer.

However, this is not the only hindrance: the operating machines find it difficult to enter and work in most hilly grounds, because of the steep slopes and the small size of plots where mechanisation is both difficult and costly.

Other problems linked to mechanical tillings which still await to be solved regard:
a) erosion hazards and stock wounds which pave the way to the onset of lignivore parasites;
b) the economic aspect in that vineyard-soil management through tillings is the dearest of all available techniques in Italy.

A major aspect - closely connected to the use of machines - must not be ruled out; this concerns the gradual but progressive ageing of farmers as a social class, leading to fewer skilled workers able to operate the machines being available. This social aspect changes the mechanisation process, putting more enphasis on machines which can either replace man or make his work much easier by dwindling down the number of more traditional jobs.

Weed cover

In a few areas sufficient rainfall in Spring and Summer enables forms of weed-cover - especially spontaneous ones - to be used.

Such practice involves several positive aspects such as improvement of soil texture and structure, of microbial life, of the soil bearing capacity, the proportion of organic matter and a reduction in soil erosion; however, it often causes a "water stress" in Summer and, at least in the first few years, it requires a greater amount of fertilizers than tilled soils do.

This is why in Italy, where emergency irrigation is not possible and there are small areas where Spring and Summer rainfall is higher than 700/800 mm - such practice is not advisable. There are, therefore, integrated solutions which enable weed-cover use to be extended to a wider environmental range.

Weeding

In order to overcome both technical and environmental problems relating to the previous systems and to reduce the economic burden, the use of weed-killing techniques has led to alternative solutions.

A few aspects, however, still await to be explained. Namely, it is a question of setting well-defined weeding patterns which enable the present problems, due to improper use, to be overcome.

The problems are temporary effectiveness, uncertain outcome, inversion flora, the danger of residue accumulation in the soil with possible adverse effects on vineyards. Barren no-tillage in soils with a high clay content has caused changes in the clay plates so as to lead to great laminar erosion. Furthermore, in this kind of ground the barren soil in Winter can cause the operating machines to slip; therefore, despite the fact that the soil bearing capacity has improved, this problem has not been solved yet.

Furthermore, barren no-tillage requires thorough knowledge of herbicides and their mode of action, of weeds and their biology so as to be correctly applied to a plastic, flexible environment such as the Italian one. This practice, which is entirely new for Italy, has proved highly successful although - alongside the positive aspects - a few flaws have shown up.

PROPOSAL

Integrated zero-tillage (weed-management concept)

In order to overcome the various difficulties which all the above-said soil management techniques involve, a few integrated solutions are being tested and some of them have begun to be applied.

These new techniques are more flexible and, therefore, can easily adapt to Italy's ever changing situations as well as being easier for winegrowers to use and taking environmental needs into greater consideration.

Underlying these new proposals, there are economic, social and technical needs which have not been met by resorting to traditional tillings; there is also a deeper knowledge of the competitive relationship between vineyards and weeds, a through awareness of the technical means available (herbicides) without spoiling the environment and an increased development of research work in this field. Furthermore, this new strategy for weed-killers use is indeed possible thanks to better knowledge of the competitiveness thresholds between weeds and vineyards.

This is why, for comparatively long periods of time, when the weed competition against vineyards is virtually non-existent (Autumn-Winter in the South, until late Spring in the North) their presence can be tolerated.

This attitude revises the approach to the problem followed up-to-now, namely the use of pre-emergence product doses such as to control the weeds all the year round.

This new proposal - based on the "threshold level concept" and the "non-competitiveness" of the weeds in the Autumn/Spring period - is an integrated system which, after having assessed the risk-benefit aspect of spontaneous weeds, involves a period of natural (or, in some cases, artificial) cover weed during cold and damp periods followed by a period in which the soil is weed-free so as to avoid competitiveness and make the most out of the vineyard's productive capacity.

This new strategy is based on a post-emergence operation coupled with a small dose of residue herbicides in order to be effective throughout the Summer.

This formula allows great adaptability to environmental conditions by, for example, extending the presence of weeds in the vineyard to periods of heavy rainfall as well.

Thanks to this approach, use of under-row weeding is also reduced; this is the first step Italian farmers usually take towards weeding (risk-threshold).

Integrated no-tillage is also the basis for a more up-to-date vision of chemical interventions, which is in line with the concept of soil management rather than with a short-sighted, limited vision of weed-control as it used to be.

THE FUTURE

"Integrated zero-tillage" is in line with the "conservation-tillage" or "reduction-tillage" which is becoming more and more popular all over the farming world since it takes into due account the relationship between herbicides, soil, vineyard and environment.

The future of weeding is linked to molecules which can meet these requirements: that is, new herbicides to support or replace triazines, herbicides that can perform positive selections of the flora, or growth regulators which enable the turf to be kept in a non-competitive condition (chemical mowing); it will therefore be up to genetic enginee-ring to produce dwarf plants so as to carry out cover crop which is, by no means, competitive.

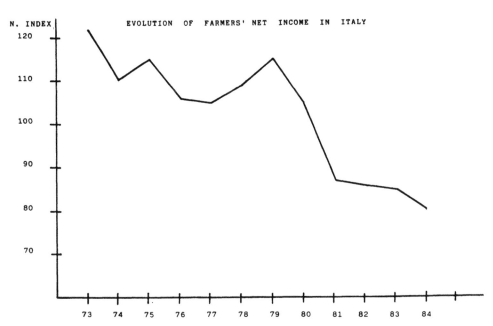

Fig. 1   The net income of vineyard growers is affected by several adverse facts, they must turn towards techniques based on less input in the product process.

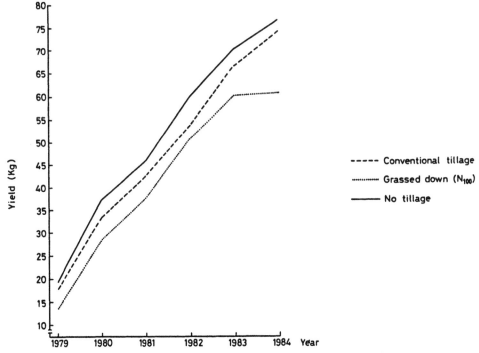

Fig. 2   Cumulative yield of "Cortese" grapes during the period 1979-1984 in plots with traditional tillage, perennial grassing, and "naked no tillage" (Az. Prago - S. Maria della Versa)

19

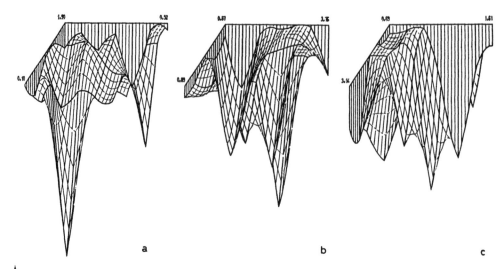

a   b   c

Fig. 3 Root development is greater in weeded soils than in grassed or tilled ones. Also root distribution along the soil profile is changed. In weeded soils the roots are more superficial (c)

Fig. 4 Soil keeping techniques apprecialy modify moisture and temperature conditions. "Naked no tillage" has generally the highest temperature and moisture content.

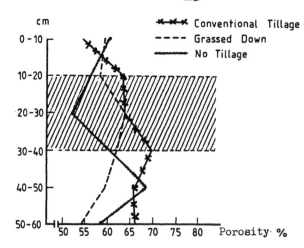

Fig. 5 Trampling causes in clayish tilled soils a considerable reduction of porosity. Grassing or "naked no tillage" improves the bearing capacity of the soil and lessens the damage from heavy equipment

SOURCE OF VARIATION	D F	BUNCH/PLANT		SUGAR %		YIELD/PLANT (Kg)		TITR. ACIDITY %O		p H		BUDS/PLANT	
		MEAN SQUARE	F	MEAN SQUARE	F	MEAN SQUARE	F	MEAN SQUARE	F	MEAN SQUARE	F	MEAN SQUARE	F
MAIN EFFECTS	3	309.751	695ns	4.764	6.481***	53.075	2.274ns	093	.550 ns	.041	6.158***	123.387	1.973ns
TREATMENT	2	349.395	784ns	4.292	5.839**	24.855	1.065ns	112	.661 ns	.048	7.245**	66.294	1.060ns
VINEYARD	1	211.592	475ns	6.053	8.234**	107.218	4.594*	057	.334 ns	.026	3.928ns	232.020	3.711ns
2-WAY INTERACTIONS	2	873.452	1.959	.035	.048	255.843	10.961***	.058	.341 ns	.007	1.123	18.073	.289ns
TREAT. x VINEYARD	2	873.452	1.959ns	.035	.048ns	255.843	10.961***	.058	.341 ns	.007	1.122ns	18.073	.289ns
EXPLAINED	5	535.231	1.200ns	2.872	3.908**	134.182	5.749	.079	466 ns	.028	4.144*	81.261	1.300ns
RESIDUAL	51	445.913		.735		23.341		.169		.007		62.523	
TOTAL	56	453.888		.926		33.237		.161		.007		64.196	

Fig. 6 - Effects of some intervention techniques on the soil on yield and must composition of Trebbiano Toscano cultivated in Lazio (Latina).
The trial was initiated in 1982, though data are referred to 1984. Sugar percent and yield pro-stok were statistically different.
Controlled grassing reduced slightly sugar content but abated costs and improved physical and microbi-cal conditions of the soil.

21

SOURCE OF VARIATION	BUNCH/PLANT	SUGAR %	YIELD/PLANT (Kg)	TITR. ACIDITY %o	p H	BUDS/PLANT
NO TILLAGE	122.33	16.00	24.08	4.42	3.73	51.59
CONVENTIONAL TILLAGE	127.15	16.00	24.30	4.88	3.69	51.98
GRASSED DOWN	118.43	15.62	22.24	4.94	3.63	48.59
L S D	--	0.67-0.85	--	--	--	--
VINEYARD (drey)	120.44	16.37	22.10	4.90	3.66	48.60
VINEYARD (irrigated)	124.29	15.72	24.85	4.80	3.70	52.63
M D S	--	0.46-0.61	2.57-3.43	--	--	--
A V E R A G E	122.40	16.04	23.50	3.17	3.68	50.65

Fig. 7 - Effects of some intervention techniques on the soil on yield and must composition of Trebbiano Toscano cultivated in Lazio (Latina).
The trial was initiated in 1982, though data are referred to 1984. Sugar percent and yield pro-stok were statistically different. Controlled grassing reduced slightly sugar content but abated costs and improved physical and microbical conditions of the soil.

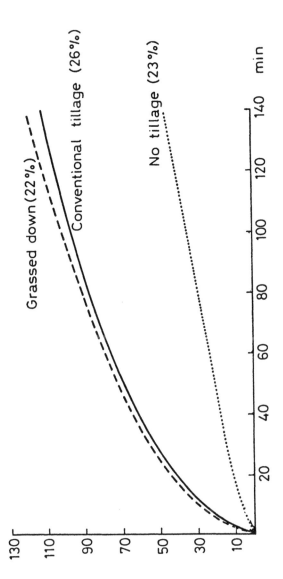

Fig. 8 - Water quantity which infiltrates the soil in time unity (1/min.) is trongly modified by intervention technique to the soil. In grassed and tilled soils water seeps through in larger quantity than in weeded ones. (Values between brackets are referred to soil moisture).

SITE OF EXPERIMENTAL FIELDS AND VINE UNDER TEST

ZERO-TILLAGE

Farm / Vine	N° M BUDS/STOCK	N°M BUNCHES/STOCK	STOCK PRODUCTION KG.M	BUNCH WEIGHT gr.M	FERTILITY	SUGAR %	ACIDITY %.	pH	WEIGHT OF PRUNING KG.
Farm: BARBERO CAVOLPI-CANELLI (AT) — MOSCATO	14,7	21,2	5,95	0,281	1,442	14,46	12,12	2,89	0,664
RETORBIDO (PV) — BARBERA	15,3	32,3	9,12	0,282	2,111	15,2	13,91	2,79	1,155
Farm: BREGA MONTU'BECCARIA — BONARDA	24,4	26,8	11,62	0,434	1,094	19,35	6,96	3,08	—
Farm: RICASOLI BROLIO (SI) — SANGIOVESE	9,5	20,4	4,85	0,238	2,147	16,55	7,96	2,91	0,791
Farm: CONSOLARO PEREZ S.BENEDETTO LUGANA — TREBBIANO DI LUGANA (ZERO TILLAGE)	28,1	46,5	17,02	0,366	1,655	14,65	10,41	3,04	1,746
Farm: AGRONOMIC TECHICAL INSTITUTE SIENA — TREBBIANO TOSCANO (ZERO TILLAGE)	23,2	36,5	8,35	0,229	1,573	15,74	9,67	2,92	1,539
Farm: BORLETTI PADOVA — MERLOT	42,7	117,5	19,65	0,167	2,752	15,61	9,02	3,12	1,691
Farm: POLA FERRARA — TREBBIANO	73,4	132,5	32,68	0,247	1,805	18,87	12,84	2,99	6,364

TILLED

Farm / Vine	N°M BUDS/STOCK	N°M BUNCHES/STOCK	STOCK PRODUCTION KG.M.	BUNCH WEIGHT gr.M	FERTILITY	SUGAR %	ACIDITY %.	pH	WEIGHT OF PRUNING KG.
Farm: BARBERO CAVOLPI-CANELLI (AT) — MOSCATO	13,2	18,0	5,47	0,364		13,85	12,04	2,91	0,659
RETORBIDO (PV) — BARBERA	14,4	28,7	7,47	0,260	1,979	15,73	13,75	2,77	1,046
Farm: BREGA MONTU'BECCARIA — BONARDA	25,4	28,8	10,74	0,373	1,134	18,87	6,77	3,10	—
Farm: RICASOLI BROLIO (SI) — SANGIOVESE	8,9	16,6	3,14	0,189	1,887	18,95	8,20	2,93	0,827
Farm: CONSOLARO PEREZ S.BENEDETTO LUGANA — TREBBIANO DI LUGANA	29,3	39,2	14,38	0,366	1,338	15,38	9,92	3,07	3,017
Farm: AGRONOMIC TECHICAL INSTITUTE SIENA — TREBBIANO TOSCANO (WEED COVER)	20,6	36,2	8,74	0,241	1,757	14,85	9,94	2,78	
Farm: BORLETTI PADOVA — MERLOT	42,7	94,4	14,02	0,148	2,211	17,78	7,85	3,16	
Farm: POLA FERRARA — TREBBIANO	71	122,5	30,43	0,248	1,725	19,07	9,13	2,95	

Fig. 9 – Effect of different techniques of soil management on different parameters (yeld-quality) in 8 vineyards growing in differents pedoclimatics areas.

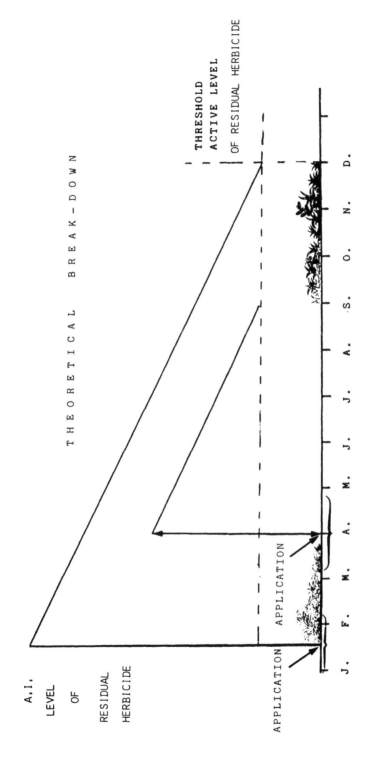

Fig. 10 The traditional barren no-tillage is based on pre-emergence application in winter, with a high level of residual herbicide, to to keep the soil clean all year round. The new strategy cuts by half the rate of residual herbicide being the objective more limited, time of application will depend on weather, soil conditions, weed development.

Chemical, cultural and biological control of *Oxalis pes-caprae* in vineyards in Greece

E.A.Paspatis
'Benaki' Phytopathological Institute, Weed Department, Kiphissia, Greece

Summary

Oxalis pes-caprae L. is a common weed in vineyards in many areas of Greece. The biological cycle of this weed and the chemical and cultural methods used for its control are presented in this paper. *Orobanche* spp. has been found parasitising on *O. pes-caprae*; so the possibility to use it as a biocontrol agent is considered. The advantages and disadvantages of the various control methods are discussed.

1. Introduction

Oxalis pes-caprae is a common weed in vineyards of Southern Greece, Crete and other islands. It is a dicotyledonous plant of the family Oxalidaceae, order Geraniales (5). This species is indigenous of South Africa and was introduced to Mediterranean countries at the beginning of the 19th century probably as an ornamental plant (2). It grows abundantly in a wide range of soil types and apart from vineyards it is a weed problem in citrus orchards and olive groves . In the case of olive groves oxalis impedes the hand-picking of olives from the ground (the usual way of olive picking in Greece) (1). Also its leaves and stems, which are picked together with olives, increase the acidity of the olive oil. Moreover many cases of animal poisoning have been reported, especially of sheep fed on oxalis, which were due to the oxalate content of the plants interfering with the calcium metabolism in the animals (4).

2. Biological cycle of Oxalis in Greece

In Greece and other Mediterranean countries oxalis does not produce any seeds and is vegetatively propagated by bulbs (2). During one growing season each plant of oxalis produces in the ground 10-40 small elongated bulbs, depending mainly on the climatic conditions. These daughter bulbs are widely dispersed by soil cultivation and transfer of soil. Also the root contraction typically demonstrated by oxalis roots results in a natural way of bulb dispersion. By this way the newly formed bulbs may be moved up to 40 cm from the parent bulb, resulting in few years time in the complete coverage of the ground by a thick carpet of oxalis vegetation.

The emergence of oxalis begins around the middle of October before the first autumn rains. The bulbs are unable to grow earlier because of their dormancy. The break of bulb dormancy is directly correlated to soil temperature and its fluctuations (3).

Bulbs of oxalis can be found in a depth of 2-50 cm or even deeper and their ability to reach the soil surface is related to the soil type. In light soils oxalis can emerge from a depth of approximately 80 cm.

Flowering begins in early February and continues until April, depending on climatic conditions and location, and it coincides with contractile root initiation. At the end of flowering the flow of photo-synthates is diverted to the underground parts of the weed so that the new bulbs can be formed. About this time, root contraction starts and the leaf development begins to decline. By the end of April the aerial parts of the plant dry out and the formation of the new bulbs is complete.

3. Oxalis as a weed problem

Oxalis is a winter weed and it could not be considered as a serious weed problem in vineyards in Greece. However, it can disturb the utiliza-tion of the fertilizers applied in the winter and the effectiveness of residual herbicides if they are applied without previous removal or destruction of oxalis vegetation. The results of incorrect application of residual herbicides will show later in the spring after the decline of oxalis, when the various annual and perrenial weeds begin to emerge.

However, there are some reservations about the eradication of oxalis in some cases, when plantations are located on slopes of poor soil. The coverage of such places by a thick carpet of oxalis in the winter protects them from erosion which is a serious problem in Greece since ancient times. Also the foliage and root debris of the weed provide the soil with some organic matter. The numerous underground tunnels opened by contractile roots of oxalis contribute to the aeration of the soil and help to retain more water from the few spring rains in Southern Greece.

Another advantage of the presence of oxalis according to our own observations in vineyards as well as in citrus orchards and olive groves, is that where oxalis is present the emergence and development of other more difficult-to-control weeds, like *Parietaria* sp. (recently a serious weed problem in Southern Greece) and other weeds of the genera *Amaranthus, Chenopodium* etc., is inhibited.

For the above reasons oxalis is considered by most vine growers in Greece and in other countries as a harmless or even useful weed (4).

4. Chemical control of oxalis

Oxyfluorfen gives good results for controlling oxalis in vineyards as well as in citrus orchards and olive groves when applied a little before or soon after the emergence of this weed in the autumn. The rate recommended for this purpose is 1200 gr ai/ha. At the end of the season, a small number of oxalis plants will regrow.

Glyphosate also gives good results when oxalis plants are taller than 10-15 cm. The recommended rate for this application is 2850 gr ai/ha. However, a large number of oxalis plants regrow in the spring after this early application of glyphosate. This results in a prolonged biological cycle of the weed which causes a water antagonism with the vines at a critical time for the growth of the plantation. Eradication of oxalis in vineyards is obtained by application of the aforementioned rate of glypho-sate when the weed is at the flowering stage, or shortly afterwards when the flow of photosynthates has been diverted to the underground parts of the weed for the formation of bulbs.

In preliminary pot and field experiments the mixture of oxyfluorfen + simazine at the rates of 950 + 3000 gr ai/ha respectively, gave good results for controlling oxalis when it was applied soon after the emergence of the weed in the autumn. Oxyfluorfen and oxyfluorfen + simazine do not significantly inhibit the sprouting of oxalis bulbs but restrain growth and development of oxalis plants.

In experiments carried out in Greece, chlorthiamid and dichlobenil in granular formulations, with or without incorporation into the soil, both at the rate of 7500 gr ai/ha, gave sufficient control of oxalis (1).

5. Cultural control of oxalis

In most Greek vineyards control of oxalis and other weeds is obtained by cultivating techniques, mainly with rotary cultivators, in the spring after the winter rains. Herbicides are used to delay the growth of weeds until the soil is ready for cultivation. As a result of such practices, oxalis has not been a serious problem in Greek vineyards. At the time of soil cultivation, i.e. in March, the growth of aerial parts of oxalis has been completed and the weed is at the stage of underground bud differentiation to propagating bulbs. With soil cultivation the underground shoots of oxalis are destroyed and the formation of bulbs is interrupted.

6. Biological control of oxalis

In recent years a very interesting phenomenon was observed in oxalis infested vineyards, citrus orchards and olive groves of Crete and Northern Peloponnese; the parasitism of oxalis by broomrape (*Orobanche* sp.). The broomrape emerges when oxalis is at the flowering stage and in most areas in a great density. This phanerogamous parasite, which lacks a rooting system or chlorophyll, is attached by haustoria on a feeding root of oxalis and is fed exclusively from the host. Nutrients necessary for the development of oxalis bulbs, formed at this time, are used by the parasite. As a result only a few small bulbs are formed, which eventually produce a poor vegetation of oxalis in the autumn. The oxalis plant population in vineyards and other plantations infested with broomrape remains low. So the weed is under control and it does not cause serious problems, but even this coverage provides adequate protection to the soil against erosion.

The possibility of using broomrape as a biocontrol agent against oxalis must be investigated. Some experiments are now being carried out for this purpose. The objectives of these experiments are the implementation of a method of infestation of vineyards with broomrape the ability of broomrape to parasitise other desirable plants adjacent to the vineyards. There is the risk that broomrape may infest sensitive crops and cause unpredictable damages. This is a potential disadvantage of the biological control of oxalis by broomrape.

The chemical as well as the cultural control of oxalis also have many disadvantages as the destruction of oxalis vegetation especially in slopes results in their erosion and the dispersion of some other difficult-to-control weeds like *Cynodon dactylon*, *Parietaria* sp. etc.

Efforts will be made towards an integrated weed control system to combine chemical, cultural and biological control methods where oxalis is a problem in vineyards, in order to minimise the possible disadvantages and increase the gains from its control.

REFERENCES

1. DAMANAKIS, M. (1976). Control of *Oxalis pes-caprae* L. with pre-emergence and post-emergence treatments. Proceedings 1976 British Crop Protection Conference - Weeds, p. 321-327.
2. GALIL, J. (1968). Vegetative dispersal in *Oxalis cernua*. American Journal of Botany 55(1) : 68-73.
3. JORDAN, L.S. and B.E. DAY (1967). Effect of temperature on growth of *Oxalis cernua* Thunb. Weeds, 15 : 285
4. PEIRCE, J.R. (1981). Beating soursob. Journal of Agriculture Western Australia, 22 (3) : 93-97.
5. YOUNG, D.P., (1968). *Oxalis* L. in Flora europaea Vol 2 p.192-193. Tutin, T.G. (Edt).

Control of chloroplastic resistant weed biotypes in vineyard: The phytotoxicity of pendimethalin

E.Beuret

Station Fédérale de Recherches Agronomiques de Changins, Nyon, Switzerland

Summary

The use of pendimethalin in vineyard can be the cause of important damages on the young expanding leaves in early vegetation stages. The lamina curls down and presents necrotic margin and oil-like spots ; the mean leaf surface is also seriously affected. It was demonstrated that the damage is caused by the vapour phase of the product, and depends on the moisture of the soil.

1. Introduction

Since 1978 the number of weeds presenting the chloroplastic resistance has increased dramatically in all the situations where triazines are used intensively : maize, vineyard, orchards, asparagus and railroads. At the moment, 13 species have been recognised as resistant with the fluorescence test developed by DUCRUET and GASQUEZ (2) : 6 of them are present in vineyards, and 7 in orchards (BEURET et NEURY, 1).

In the region of Valais, the most troublesome resistant weeds are Amaranthus retroflexus and A. lividus. As previous observations had proved that pendimethalin is very effective against these species as well as against Setaria and Digitaria which are also present in these situations, although not chloroplastic resistant, we made a number of field trials with this herbicide to precise some agronomic parameters like effectiveness, persistance of action, toxicity, etc. After two years without any damage, and a good control of the target weeds, we observed on the third year an important phytotoxicity for the vine plants. As we suspected a toxicity through a phenomenon of volatility, we made some experiments to verify this supposition.

2. Material and methods

In all experiments, pendimethalin was used as formulated by Cyanamid Company under the commercial name of Stomp which contains 330 g/1 pendimethalin.

2.1. Experiments without soil-herbicide contact

Young vine plants cultivated in pots were placed at the beginning of leaf production (stage E - G) in contact with air containing vapour of pendimethalin, in two different ways. In the first experiment two plants (one Gamay and one Chasselas) were placed in a box on whose bottom a solu-

31

tion of pendimethalin $9.4 \cdot 10^{-3}$M (0.8 % Stomp) was poured. The pots were placed on supports to avoid any contact with the solution. The surface of the box was 0.2 m^2. The box was then covered with a transparent plastic bag carefully sealed, and placed in glasshouses at 3 different temperatures : 12°C - 20°C and 28°C ; the plants were kept in this situation during 24 h ; after this time, they were taken off from the boxes and kept free of pendimethalin during a fortnight in a glasshouse at 20°C for assessment. Each treatment was made in 3 replicates.

In the second experiment, four pots (two Gamay and two Chasselas) were put into a box through which a stream of air charged with pendimethalin was blown. The enrichment with pendimethalin was performed in placing before the inlet pipe a gas washing bottle in which a solution of pendimethalin $9.4 \cdot 10^{-3}$M (0.8 % Stomp) had been poured. The plants stayed in this situation at 22°C during 24 h, after which they were kept a fortnight in a glasshouse at 20°C for assessment. The volume of the box was 1.60 m^3 and two replicates were made.

2.2. Experiment with soil-herbicide contact

In six little compartments of a glasshouse (volume 55 m^3), soil (a sandy clay loam soil) was spread on the tables in a sheet of 8 cm and a surface of 6.6 m^2. On this earth, pendimethalin was sprayed at a rate of 2640 g/ha (8 1 Stomp). After application, 12 vine plants (6 Gamay and 6 Chasselas) grown in pots and being at the beginning of leaf production (Stage E - G) were placed on the earth. The treated soils was maintained at three different moisture level : in one system (N° 1), no water was added so that the soil was maintained dry ; in the second (N° 2), the earth received 20 mm water before the pots were put on it ; in the third treatment (N° 3), the earth was watered three times with 5 mm, at 7 days intervals, the pots being put on the soil after the first watering. The plants stayed 4 weeks in this situation in a temperature oscillating between 20 and 25°C. At the end of this period, the leaves were cut and pressed, and the surface was determined. The experiment was made in two replicates.

3. Results and discussions

3.1. Experiments without soil-herbicide contact

In neither of the plants of the two experiments, we were able to observe any phytotoxic symptom on the foliage of the treated plants. Moreover, the mean leaf surface of the treated plants did not differ from the one of the check plants (Tables I and II).

3.2. Experiment with soil-herbicide contact

Already three days after treatment N° 3, the leaves of Chasselas showed precise symptoms of toxicity : the same symptoms appeared on Gamay only two days later. In both cultivars, the symptoms were the same : the leaf margin was ± necrotic and irregular with a certain tendency to curl downward. The lamina presented clear oil-like spots. In all cases where pendimethalin had been sprayed, the symptoms occured, and the mean leaf surface was seriously affected (Table III). The most damaging treat-

32

Table I.

Mean leaf surface of vine plants in contact with pendimethalin vapour without soil. First experiment : pendimethalin $9.4 \cdot 10^{-3}$M spread on the bottom of the box. (see text).
10 leaves/plant   -   3 plants/treatment.

	Surface (cm^2)			
Temperature	Gamay		Chasselas	
	Treated	Check	Treated	Check
12°C	20.8	21.8	15.0	17.3
20°C	18.3	20.0	17.2	16.6
28°C	22.1	21.7	17.8	16.2

Table II.

Mean leaf surface of vine plants in contact with pendimethalin vapour without soil. Second experiment : pendimethalin $9.4 \cdot 10^{-3}$M blown with gas washing bottle (see text).
10 leaves/plant   -   4 plants/treatment.

	Surface (cm^2)	
	Gamay	Chasselas
Treated	21.3	18.9
Check	20.6	21.2

Table III.

Mean leaf surface of vine plants in contact with pendimethalin vapour in presence of soil (see text).
20 leaves/plant   -   6 plants/treatment

	Gamay				Chasselas			
	Check	No 1	No 2	No 3	Check	No 1	No 2	No 3
Surface (cm^2)	47.36	27.14	20.29	9.94	36.24	21.02	14.65	10.18

ment was N° 3 where the soil was watered three times and the less severe was treatment N° 1 without soil watering. Between these two extremes was treatment N° 2 with only one watering.

The volatility of a compound, that is to say its intrinsic tendency to evaporate, is usually indicated by its vapour pressure. Dinitroaniline herbicide may be lost by volatility as is known for trifluralin which must be quickly incorporated to prevent this loss (FRYER and MAKEPEACE, 3). Vapour pressure is an equilibrium property, but the atmosphere can be regarded as an infinite sink so that volatilisation of pesticides is a dynamic process in which kinetics are important. Hence the evaporation rate of a compound applied to an inert surface depends not only on its vapour pressure and the temperature, but also on air flow and the ratio of the surface area to applied mass of the deposit (HANCE, 4). This could be the explanation why in the first two experiments without soil-herbicide contact, no phytotoxicity could be observered.

The important phytotoxicity of pendimethalin observed in treatments on soil, and the increase of efficiency with water content of the soil is a consequence of the adsorbtion and air-water partition of the compound. Dinitroanilines are active in the vapour phase, so as water content increases, moisture competes for adsorption sites and less herbicide is adsorbed, resulting in a greater activity. Such observations were made by OKAFOR et al. (6) with dinitramine, and by HOLLINGSWORTH (5) with trifluralin. As far as we know the use of pendimethalin till now had never conducted to such an observation.

On the field the most important toxicity was observed after light but frequent rainfalls, that is not the usual climate of Valais. Nevertheless, although such accidents are rather unlikely, the risk is too high to accept the clearance of such a product in vineyard, because when present the damage can lead to a complete loss of the production.

On the other hand, care must be taken when pendimethalin is to be used too close to a vineyard. In one case, we could observe damages on vine plants after treating an adjacent carrot crop with pendimethalin. The toxicity of pendimethalin in such a situation does not mean that all dinitroaniline herbicides are not acceptable in vineyard. We have recently tested oryzalin in a similar experiment with soil and different moisture content, without observing any toxicity one vine plants.

References

1. BEURET,E. et NEURY,G. (1983). Les adventices résistant à l'action des triazines dans les vignobles de Suisse romande.
   12ème Conférence du COLUMA, 2, 301-309.

2. DUCRUET, J.M. et GASQUEZ, J. (1978). Observation de la fluorescence sur feuille entière et mise en évidence de la résistance chloroplastique à l'atrazine chez Chenopodium album L. et Poa annua L.
   Chemosphère 8, 691-696.

3. FRYER, J.D. and MAKEPEACE, R.J. (1978). Weed control handbook.
   Vol. II Recommendations. 8th Edition, 532 pp.

4. HANCE, R.J. (1980). Transport in the vapour phase. in Interactions between herbicides and the soil.
   R.J. Hance ed. Academic Press London, 59-81.

5. HOLLINGSWORTH, E.B. (1980). Volatility of trifluralin from field soil. Weed Science 28 (2), 224-228.

6. OKAFOR, L.I., SAGAR, G.R. and SHORROCKS, U.M. (1983). Biological activity of dinitramine in soils. II. Soil organic matter and soil moisture content. Weed Research 23, 199-206.

Experiments on weed control in vines

G.Marocchi

Osservatorio Regionale per le Malattie delle Piante, Bologna, Italy

Summary

We report the good results of a trial which we began 7 years ago. Now in the vineyard there is an excellent total weed control that we obtained using simazine, paraquat, glyphosate and so on. Besides we observed better chemical and physical conditions of the soil. The chemical weed control was not so expensive as other agricultural systems.

Weed control in the vineyard is of much interest in Italy but the practical knowledge of the so-named "non-cultivation" is less known than in other countries, because a number of reasons have hindered the dissemination of information. These reasons can be summrized as follows:-

(a) Lack of knowledge of the techniques

It is evident that one can have a resistance against such a radical change. But however, the obstacle has been overcome by many trials already started or in progress. Among these is a trial started 8 years ago and still being followed up by collecting data, remarks, observations and other information.

(b) Fear of causing damage

The trials have demonstrated that there are no dangers and that instead there were crop benefits to be obtained.

(c) Attachment to old traditions

There exists a resistance by the farmers not to abandon the working of the soil. But foreign experience and the results of many trials are demonstrating that it is possible to change the system.

(d) No evaluation of the damage caused by weeds

A difficult valuation, but many figures indicate that weeds can damage vineyards by 20-30-50% and more.

(e) Cost of herbicides

This pretext is difficult to justify because of the principal reason that induces a grower to use chemicals is exactly that of saving costs, because a rational usage of herbicides makes the management of the vineyard less costly. It is evident that this technique was started by leading firms and by the biggest and the better managed.

To describe this new and interesting technique there exists already many components and results of our Italian experiments and part of these is the trial mentioned. But there are also figures, remarks and observations derived by big companies which are already using the practice. All the positive results on relevant matters are being examined and help to make a useful comparison between our situation and that in France. Because from the French much information is already available as they started their experiments on non-cultivation 20 years ago and in many areas over 80% of their vineyards are managed in this way.

The technique of 'non-cultivation' is very easy and generally requires two types of treatment. One during the winter time, until February/March with products which are principally soil-acting and one (or more than one) with leaf-acting herbicides. During the winter period simazine is often used as also is the mixture terbutilazin+terbumeton (Caragard). Also in France and in Italy growers started with rather high doses of these residuals, with lower doses in subsequent years. In every case it is correct not to exceed recommended doses, not for fear of causing damage but because high doses are very expensive.

The trial, the results of which we now examine, was started in 1978 and from that time on weed control was by chemicals only, alternating occasionally winter applications with summer operations.

The remarks and the observations made during these 8 years of the trial are summarized in this way:

(a) <u>Appearance of the cultivation and production of grapes in quantity and quality.</u> The vineyard is now in optimal conditions and with a vegetative vigour exactly equal to the cultivation with traditional systems. There are no visual sign of injury and the production of grapes and also the quality were equal in the 'non-cultivated' section and in that mechanically weed controlled. In addition as there was no costly "working" of the soil, the risk of damage to branches and roots was avoided.

(b) <u>Condition of the soil.</u> Stratigraphic examinations were made at various soil depths and an advantage of the new technique was the remarkable improvement of the physical condition of the soil, like structure, permeability and life of micro-organisms. The root-system of the vine plants, which in "worked" ground is destroyed in the superficial strata of the soil in a thickness of 20-30 cm is uniformly distributed throughout the surface of the 'non-cultivation' system. This is a notable advantage for mineral nutrition and also for resistance to drought. Figures elaborated from the current trials, from observations made in other applications in Italy, but especially confirmed by the figures of the French experiences, show us that in very dry years especially the soils of the 'non-cultivation' system have a higher moisture content. It is also easy to understand that with a dense root-system in the soil surface, the vineyard also benefits immediately from rainfalls of only a few millimeters.

(c) <u>Traversability of the soils and resistance against erosion</u> The trial described is situated in the lowland and therefore the problem of erosion does not exist. In contrast this is a well-known disadvantage in vineyards on hillsides. It can be said with certainty that 'non-cultivation' brings an undeniable benefit, because it is in the "worked" soils that the water makes the biggest erosion impact and transports the soil downwards. In the 'non-cultivation' trial was started distinctly a better "lift" of the soil than in the "worked" soil during the harvest of the grapes or the treatments with different machines. This fact caused a few surprises at first as it was an apparently contradictory matter: the soil if also more soft, porous and easy to work with spade and hoe, is anyhow more difficult to drive over with machines or other equipment.

(d) Fertilization. There are always differences of opinion on the best way to proceed when fertilisation is discussed. In the current trial this problem was solved by the distribution of organic substance on the surface and without covering with earth: in a short time there has been the decomposition with a major benefit to the cultivation, both for the appearance of the earth and for the nutrition of the vineplants.

(e) The economic appearance. This is the most important element for the adoption of this technique. The French figures on this matter are quite clear. It is obvious that the calculations and the necessary comparisons must be made over a rather long period of time, before the use of herbicides can be reduced to any extent. Manual or mechanical operatins (in a vineyard managed by traditional systems) will always have to be done every year and it is not feasible to reduce them. If you think also about the continuously increasing costs, it is easy to imagine that the differences in cost between chemical and mechanical weed control will increase further, considering also the trend for the cost of fuel to rise more rapidly than the cost of herbicides. For the current trial, schematized at a maximum, it can be calculated - in the average of 8 years - a medium cost of Lire 280.000 per year in the 'non-cultivated' compared with 450.000 in the "worked" soils. Already now, without any doubt, we can see the net convenience. It is obvious that this cost difference will increase further in the next few years, when the costs for chemical weed control will be reduced relative to the constantly high and more rapidly rising cost of mechanical work.

It seems, that there are good reasons to believe that the 'non-cultivation' system has already surpassed every test. Anyhow this conclusions is reached from some basic concepts, the results and the advantages are certain. The French experience, the figures of our current trials, the undiscussed expertise and the spirit of observation of our growers and the possibility to count on a number of excellent chemical products (simazine,Caragard, diuron, desiccants, Roundup and others) are all sure guarantees of success of this technique also in Italy.

Session 2
Weed problems in soft fruits

Chairman: D.Seipp

Volunteer crops as weeds in soft fruit plantations

H.M.Lawson

Scottish Crop Research Institute, Dundee, UK

Summary

Problems caused by volunteer crops in soft fruit plantations
are outlined and data presented on their ecology and control.
Species discussed include cereals, potato, oilseed rape,
field bean, flowerbulbs, pasture grasses, white clover and
soft fruit seedlings. They are resistant to many of the
herbicides normally used in fruit plantations and frequently
involve the grower in additional expenditure of time and money
on control methods. By acting as a bridge for pests and
diseases, volunteer crops reduce the benefits of crop rotation.
Seedlings of the same species in a soft fruit plantation are
particularly difficult to control.

Introduction

With the wide range of herbicides now available for use in soft fruit
plantations, chemical control of most common annual and perennial weed
species is now feasible, although not always practicable. In recent
years, much attention has been focussed on annual species which show
indications of developing resistance to the most commonly-used herbicides,
or on hitherto minor perennial species for which no effective remedy as
yet exists. Little information is available on the incidence and control
of a less-obvious group of weeds i.e. volunteer crops. With the exception
of cereals, they are seldom mentioned in advisory publications on weed
control in soft fruit crops and are rarely included in manufacturers'
weed lists on relevant herbicide labels. In reports on pesticide usage
surveys on soft fruit crops carried out by the Ministry of Agriculture,
Fisheries and Food in England and Wales and by the Department of
Agriculture and Fisheries for Scotland, farmers have on occasion identified volunteer crops as unsolved problems or as the reason for specific
herbicide treatments. Judging by the numbers of enquiries reaching SCRI
from horticultural advisers, volunteer crops are of considerably more
importance in soft fruit than either 'resistant' annual or minor perennial
weeds. SCRI has an ongoing project on the ecology and control of
volunteer crops in rotations and this paper summarises the information
gathered to date as it relates to soft fruit plantations.

Cereals

At Invergowrie, as on many farms, soft fruit crops normally follow cereals in the rotation. Over the years, the removal by hand of volunteer barley from herbicide evaluation plots has been a constant chore in maiden crops. In Eastern Scotland, spring planting of fruit is often associated with dry soil conditions and residual herbicides frequently have little or no effect on cereals. Autumn planting of fruit can also lead to problems, since recently-shed grain will germinate readily in the seed-bed conditions appropriate to newly-planted fruit. In undisturbed plantations the first flush of cereal seedlings should account for the majority of the potential problem but soil cultivation may result in further flushes by bringing buried seeds within germination depth (7-8 cm). With the strawberry crop there is a further source of potential infestation from seed in straw. In alleyways, cereal seedlings are normally controlled with paraquat. Larger plants may need more than one application. In crop rows the solution until fairly recently has been hand-hoeing or pulling, but the advent of the specific graminicide alloxydim sodium now permits removal of volunteer cereals by selective application in all the major soft fruit crops without risk of injury to crop plants. Propyzamide also controls volunteer cereals, but label restrictions on crop age, soil type, application timing and growing system greatly reduce its potential use for this purpose. Encouraging shed grain to germinate in stubbles for control with paraquat or glyphosate prior to ploughing can help to reduce the seedbank encountered by subsequent fruit crops.

Potatoes

Traditionally potatoes were often grown as a cleaning crop prior to planting soft fruit crops. It was also not unusual for growers to plant potatoes as an inter-row crop in newly-planted bush and cane fruit plantations. These practices are less common nowadays, but potato crops grown at any point earlier in the rotation can lead to problems with volunteer potatoes in young fruit plantations. Plants from tubers are particularly difficult to eliminate in fruit crops. Hand-digging or spot treatment with glyphosate in the alleys are the main control measures practised. The former is slow and expensive while the latter risks injury to the crop. The residual or selective post-emergence herbicides available for soft fruit crops are not effective, while paraquat gives only a temporary kill of top-growth. The best remedy for this problem is effective control of volunteer tubers earlier in the rotation, e.g. by pre-harvest treatment in cereals, and the avoidance of potatoes as a crop grown shortly before or in new plantations. At one trial site in Eastern Scotland where spring-planted raspberries had followed potatoes, 4-5000 volunteer potato plants/ha (from tubers) were recorded in the raspberry alleyways in late September, despite the farmer having dug and removed all visible plants twice already during the first growing season. This was however, only part of his problem; 15,000 potato seedlings/ha were also recorded in September and analysis of soil cores indicated the presence of a further 16 ($\pm$3) million true potato seeds/ha in the top 20 cm of soil. Enquiries confirmed that the previous potato crop (cv. Maris Piper) had produced large numbers of berries. Potato seedlings normally germinate from early May onwards and seedlings emerging up until late June will produce tubers before the end of the growing season. Thus, even if the farmer has taken steps to control volunteer tuber populations, a soil seedbank of true potato seed may ensure a continuing problem. There are no label recommendations for the control of potato seedlings in soft fruit crops, and experiments at SCRI have shown no useful control with the post-emergence herbicides normally used in these crops. However, spring applications of bromacil at normal rates of application for control of annual weeds in raspberry have consistently prevented later emergence of potato seedlings, while simazine, diuron, cyanazine and lenacil have given variable results, depending on soil moisture conditions. Propachlor, trifluralin, chlorthal dimethyl and pendimethalin have not proved effective in any of our investigations to date. Directed applications of paraquat or dinoseb-in-oil in alleyways are very effective provided that they are applied before tuber initiation. Volunteer potatoes probably cause more problems in commercial soft fruit crops than any of the other species mentioned in this paper.

Oilseed Rape

The acreage of oilseed rape is increasing all over the country and its spread to farms which include horticultural crops in their rotations is creating a major volunteer weed problem for these crops. Vegetables, especially other brassicas, are particularly at risk but fruit grown in arable rotations may also be affected. Strawberries are the most vulnerable of the fruit crops, since chlorthal dimethyl, lenacil, propachlor and simazine are ineffective at rates of application used in this crop. Problems are most likely to arise in newly-planted crops, especially if these follow oilseed rape, but the seed can survive for several years in the soil and volunteer plants may themselves be able to replenish the seedbank. On one site at Invergowrie strawberries were planted in spring 1979, three years after oilseed rape. In each of the following four years rape seedlings emerged at intervals from early spring onwards. Regular hand-hoeing was necessary to prevent flowering and seeding by these plants, but seedlings emerging after strawing in May were able to develop undisturbed and to return seed to the soil before straw-removal in autumn. No mechanical cultivation occurred after planting and the plantation received routine applications of lenacil (spring) and simazine (autumn) every year. It is not possible to say how much of the annual emergence of seedlings in later years was due to seed surviving from the original crop or to seed returned by successful seedlings. It is clear, however, that oilseed rape could readily find a niche as a weed where normal strawberry management is practised. In the other fruit crops, directed alley treatments with paraquat or dinoseb-in-oil during the maiden year are highly effective on seedlings and young plants, but we have been unable to achieve effective control with residual herbicides other than bromacil.

Field Beans

This crop does not immediately come to mind as a potential weed of soft fruit crops, but shed seed can produce large numbers of seedlings in subsequent crops. Vegetables are particularly vulnerable, but the seedlings are also difficult to control in newly-planted fruit crops. They are resistant to normal rates of the residual herbicides used in these crops and while the foliage can be killed by paraquat, recovery is rapid. Hand-hoeing is also relatively ineffective, unless repeated several times.

Flowerbulbs

Volunteer narcissus and tulips are also very difficult to control effectively in newly-planted fruit crops. Residual herbicides are ineffective. Paraquat gives only top-kill and selective application of glyphosate is practicable only in the alleyways. At Invergowrie volunteer tulips have produced foliage and sometimes flowers in each of the past 5 years in a non-cultivated strawberry plantation given the full range of herbicide treatment appropriate to this crop. The volunteer plants have been kept in check, but not killed, by hand-hoeing in May prior to strawing. They produce no further growth in that year, but have obviously accumulated sufficient reserves by May to survive until the following spring. Narcissus behave similarly. Hand-digging is the only fully-effective remedy.

Pasture Species

Soft fruit crops can be grown in rotations which include leys and may even follow them directly. Clumps of cocksfoot, timothy or perennial ryegrass are not uncommon in fruit plantations and are difficult to control in the crop row. Residual herbicides are in general not very effective; propyzamide is effective on all except cocksfoot, but cannot be used in the first growing season, when the weeds are most sensitive. Alloxydim sodium is claimed to control ryegrass seedlings selectively in soft fruit crops.

White clover is a particularly difficult problem in strawberry rows where, as well as competing with the crop, it is an alternate host for strawberry green petal mycoplasma. The only treatment recommended for established plants in strawberry is with ethofumesate in cv. Cambridge Favourite during late autumn/early winter. Other cultivars and other dates are excluded for reasons of crop safety. Residual herbicides have prevented the emergence of clover seedlings in some seasons and not in others in our trials, depending probably on soil moisture conditions at and after herbicide application. Clover seed can also survive for many years in the soil before germinating, so that crops planted several years after pasture may be affected and seedlings can appear in long-established fruit plantations. In bush and cane fruits dichlobenil will give control of established white clover plants at high rates of application, but other residual herbicides are generally ineffective.

Fruit Seedlings

Soft fruit crops can leave very large numbers of seeds in the soil to offer potential problems to succeeding crops. Soil cores taken from a four year old plantation of strawberry cv. Cambridge Favourite produced the equivalent of 240 ($\pm$32) million seeds/ha in the top 20 cm of soil. Similar soil samples from an 11 year-old plantation of raspberry cv. Glen Clova yielded the equivalent of 98 ($\pm$10) million seeds/ha. No records are available for black currant, but it would be reasonable to expect populations of a similar order of magnitude. Soft fruit seedlings are readily controlled by hormone-type cereal herbicides, but are highly tolerant of many residual herbicides. They have persisted in the soil seedbank for at least 5 years in SCRI trials and cause volunteer problems in subsequent fruit and vegetable crops in our rotations.

46

Substantial numbers of strawberry seedlings were recorded in a maiden strawberry crop, planted 5 years after the previous strawberry crop, and the seedlings were too numerous to suggest other than that they were derived from the seedbank left by the earlier crop. These seedlings, unless removed by hand, became incorporated into the matted row to the detriment of fruit quality and uniformity. Soil cores taken from a maiden plantation on a replant site, where raspberry followed raspberry with only a one-year gap, indicated the presence of 81 ($\pm$3) million raspberry seeds/ha (of the previous cultivar) in the top 10 cm of soil. Again there was a risk of seedlings becoming incorporated into the cropping row. Seedlings of bush and cane fruits are fairly common weeds in matted rows of strawberry. In several recorded cases, these crops had not been grown earlier in the rotation on these sites, so that the seed was probably transported by birds from adjacent fields. High populations of seedlings of black currants and raspberries are common in established plantations of these crops, as a result of unharvested fruit falling to the soil; strawberry seedlings can also be found in fruiting strawberry plantations following adverse harvesting conditions in the previous year.

In general, fruit seedlings are not controlled by the residual herbicides used in soft fruit plantations. Paraquat can kill small seedlings, but not established plants, in alleyways; hand-digging is the only effective remedy in the crop row.

The risk of a large residual seedbank of volunteer seedlings affecting propagation stocks of the same species, such as strawberry runner beds or raspberry spawn beds is probably small since these crops are usually grown well away from fruiting plantations. Nevertheless, this type of contamination, even from seeds dispersed by birds, could be costly but very necessary to control.

Other Species

Occasional reports are received of weed problems in fruit plantations resulting from earlier crops of horse radish, Jerusalem artichoke, rhubarb and, on one occasion, sunflower.

Conclusions

Other than when they occur in large numbers per unit area in soft fruit crops, volunteer crops are unlikely to result in substantial decreases in yield. If not controlled, however, they impede other aspects of plantation management including harvest. If allowed to seed or to develop perennial root systems, they will present even greater problems in the following year. Their main drawback, however, is that they require costly alterations to normal weed control programmes, such as hand-hoeing, digging, mechanical cultivation or spot-treatment with expensive herbicides. This involves increased time and expenditure on weed control which would not otherwise have been incurred and may reduce the effectiveness of other herbicide treatments e.g. by disturbing the soil surface. Additionally, volunteer crops act as a bridge for pests and diseases, reducing the benefits of rotational practice - another hidden cost. The importance of volunteer crops is therefore much greater than their actual numbers in a particular crop may suggest and their control should be regarded as a rotational expense, rather than be

47

apportioned to one particular crop. Information from investigations at SCRI into soil seedbanks of volunteer species, and on their susceptibility to herbicides used in fruit plantations, is being used to develop strategies for their control and to encourage manufacturers to include more volunteer crops in the weed lists on herbicide labels.

REFERENCES

1. ANON. (1982). Weed control in strawberries. Booklet 2255 Ministry of Agriculture, Fisheries and Food, London.

2. ANON. (1983). Weed control in bush and cane fruits. Booklet 2264 Ministry of Agriculture, Fisheries and Food, London.

3. BOWEN, H.M., CUTLER, J.R., WOOD, J. & TUCKER, G.G. (1983). Soft fruit 1980. Pesticide Usage in Scotland Survey Report 32. Department of Agriculture and Fisheries for Scotland, Edinburgh.

4. LAWSON, H.M. (1981). Potato seedlings as weeds: a new slant on the ground-keeper problem. Proceedings Crop Protection in Northern Britain 1981, pp.137-142.

5. LAWSON, H.M. & WISEMAN, J.S. (1980). Weed control in crop rotations. Volunteer crops. Annual Report Scottish Horticultural Research Institute for 1979 pp.35-36.

6. LAWSON, H.M. & WISEMAN, J.S. (1982). Effects of horticultural herbicides on volunteer arable crops. Tests of Agrochemicals and Cultivars No. 3 (Annals of Applied Biology 100, Supplement), pp.80-81.

7. LAWSON, H.M. & WISEMAN, J.S. (1984). The post-emergence activity of ten herbicides on volunteer potato seedlings. Tests of Agrochemicals and Cultivars No. 5 (Annals of Applied Biology 104, Supplement), pp.86-87.

8. SLY, J.M.A. (1982). Soft fruit 1980. Pesticide Usage Survey Report 24. Ministry of Agriculture, Fisheries and Food, London.

Weed control in currants and raspberries in Denmark

G.Noye
National Weed Research Institute, Slagelse, Denmark

Summary

The use of a herbicide in Denmark prerequires approval for its use and for the product label by the National Agency of Environmental Protection.

The soil-acting herbicides simazine, terbuthylazine, diuron and dichlobenil are, at present, approved for the control of germinating dicotyledonous weeds. Propyzamide is likewise approved, especially for the control of grass weeds. The above mentioned chemicals must be used before the fruit bushes come into flowering.

For the control of weeds, post emergent, a few foliage-applied herbicides are approved and are, as yet, permitted for use throughout the growing season, however, this necessitates screened spraying. The contact herbicides paraquat and diquat are used most frequently, both as a mixture with soil herbicides and for burning off weeds throughout the growing season.

The translocated herbicides of the phenoxy acid group can also be used throughout the growing season, as long as they are in the form of amine salts. MCPA is used the most, especially for the control of Convolvulus and Equisetum spp, but it may also be used to control Cirsium arvense. 2,4-D is occasionally used instead of MCPA especially if the weed population contains Cirsium and other members of the compositae family.

Urtica dioica and Galium spp., are normally controlled by mechlorprop, while dichlorprop is more suitable suitable for the control of Polygonum spp, especially P. amphibium.

1.1 INTRODUCTION

The use of a herbicide in Denmark prerequires approval for its use and for the product label by the National Agency of Environmental Protection.

1.2 BIOLOGICAL TESTING

A biological test is carried out at the National Weed Research Institute, if required by the chemical firm, to assess the effect of the chemical and its phytotoxicity. This test is, at present, being carried out on the newer herbicides 'Basta' (glufosinate), 'Fusilade' (fluazifop-butyl) and 'Matrigon' (chlorpyralid).

Discussed below are the chemicals which are approved for use in fruit bushes.

1.3 SOIL-APPLIED HERBICIDES

Soil-applied herbicides, which must be used before flowering, are mentioned in table 1. Simazine is the herbicide most used for the control of annual weeds preemergence. Where simazine has been used over a period of years, a weed population, which cannot be controlled by simazine, is often established; for example, Senecio vulgaris and Galium aparine. Where these two weeds present a problem, terbuthylazine, or diuron, used as a soil-applied herbicide are chosen, whereby the simazine-resistant weeds are controlled. As both terbuthylazine and diuron are more phytotoxic to the crop than simazine, it is recommended that, after using the other soil-applied herbicides for a couple of years, simazine should be used again.

Propyzamide is used for the control of grass weeds, in the late autumn, followed by a spring-time application of one of the other soil-applied herbicides. However, this is usually only necessary if simazine is to be used.

Table I

Soil applied herbicides (kg a.i.)

	Blackcurrants	Redcurrants	Raspberries
Simazine	1.0 – 3.0	1.0 – 3.0	1.0 – 2.0
Terbuthylazin	2.5 – 6.0	2.5 – 6.0	1.5 – 4.0
Propyzamide	1.5 – 2.0	1.5 – 2.0	1.5 – 2.0
Dichlobenil	4.0 – 8.0	4.0 – 8.0	4.0 – 6.0
Diuron (until 1986)	1.6 – 3.2	1.6 – 3.2	1.6

1.4 FOLIAGE-APPLIED HERBICIDES

For the control of weeds, postemergent, a few foliage applied herbicides have been approved and these are, as yet, permitted for use throughout the growing season, however, this necessitates screened spraying. The contact herbicides, paraquat and diquat, are used most frequently, both as a mixture with soil-applied herbicides and for burning off weeds throughout the growing season.

The translocated herbicides of the phenoxy acid group can also be used during the growing season, as long as they are in the form of amine salts. MCPA is used most, especially for the control of Convolvulus and Equisetum spp, but it may also be used to control Cirsium arvense. 2,4 D is occasionally used instead of MCPA, especially if the weed population contains Cirsium and other members of the compositae family. Urtica dioica and Galium spp. are normally controlled by mechloprop, whereas dichlorprop is more suitable for the control of Polygonum spp, especially P. amphibium.

Where raspberries are harvested mechanically, root suckers present a problem. These suckers are controlled by dinoseb applied when they are 15 cm high. Depending on the harvesting method, it may be neccesary to apply dinoseb twice within a year.

Table II

Folige-applied herbicides (kg a.i.)

	Blackcurrants	Redcurrants	Raspberries
Diquat	0.6	0.6	0.6
Paraquat	1.0	1.0	1.0
MCPA	1.0	1.0	1.0
Dichlorprop	2.0	2.0	2.0
Mechlorprop	2.0	2.0	2.0
2,4-D	0.8 - 1.0	0.8 - 1.0	-
Dinoseb	-	-	1.5 - 2,0

Problem weeds in soft fruit in England

A.J.Greenfield

Agricultural Development and Advisory Service, Weed Research Organisation, Oxford, UK

Summary

There are many herbicides now available with possibilities for use within soft fruit crops. In spite of this armoury, however, perhaps because of changing methods of crop growing, and changes in other agricultural practices some particular weed problems still occur. These problem weeds, and possible control methods are discussed, according to weed type. The weed species discussed have been divided into several different categories, relating to their annual or perennial nature, susceptibility or resistance to herbicides, and the possible reasons for their build up in recent years.

1.1 Introduction

In spite of a fairly large choice of herbicides with recommendations for use in soft fruit crops in England there are still many weed problems that occur to varying degrees for a variety of reasons in these crops.

Methods of growing these crops in England have changed with time, and the market outlets to which the produce has been presented have also changed. Over the past ten years the acreage of soft fruit grown for the Pick Your Own trade has increased greatly with an estimated 25 percent of the strawberry area down to PYO, with the figures for raspberries, blackcurrants and red and white currants being 30, 12.5 and 50 percent respectively (FSPA, 1982). The PYO trade demands that crops are grown in areas which are easily accessible to the general public, for whom the crop is grown. This can often result in the crops being grown in fields which are not totally suitable for growing soft fruit, with problems of weed populations or soil conditions which require solutions before embarking on the venture, but which are often overlooked for reasons of timing. Ten years ago most holdings also had labour available and would accept mechanical or chemical weed control of 80 percent with the remaining 20 percent being hand weed control. Nowadays, particularly on PYO holdings where a pool of casual labour for harvest is not available to call on for other hand work, total weed control is expected from mechanical or chemical means.

Changes in agriculture, particularly the move to minimum cultivations for cereals, which often occur in a fruit rotation, or have preceeded a farm change to PYO fruit growing, have led to an increase in the population of autumn germinating weeds such as Veronica spp. (speedwells), Viola arvensis (field pansy) and Galium aparine (cleavers), which are resistant to both simazine and lenacil and have thus become problems in many fruit holdings where these two herbicides are the basis of the weed control programme. Weed species such as Equisetum spp. (horsetails) and other deep rooted perennials have also increased as a result of seasons where cereal crop growth has been reduced and thus less competitive. However, the introduction of the translocated herbicide glyphosate, which can be used pre-

53

harvest of cereals has resulted in a very effective control of some problem perennial weeds such as <u>Convolvulus arvensis</u> (field bindweed).

1.2 <u>Current Problem Weed Types</u>

Weed problems can be divided into several categories
1.   Annual weeds normally controlled by herbicides used can some-times be controlled ineffectively as a result of conditions such as dry weather following application of the herbicide. As growers attempt to lengthen the period of weed control into the growing season by later spring application so the problems become more likely. One way to help overcome this is to apply the first application of herbicides earlier in the year and top up just prior to flowering, and again after harvest, thereby ensuring a sufficient level of herbicide in the soil for as long as possible.
2.   Annual weeds resistant to the normal weed control programmes are well known. Species such as <u>G. aparine</u> which germinate mainly in the autumn have already been mentioned. The resistance of such weeds to the herbicides used in basic programmes such as simazine and lenacil is fort-unately a surmountable problem, as some of the more recently introduced herbicides, e.g. pendimethalin will effectively control these weeds. Other weeds do seem to appear, however, when new herbicides are used, and as a result of changes in agriculture. <u>Geranium molle</u> (crane's-bill) is one such weed, and it has been suggested following screening that propachlor should give good control of this species (Jones, 1984).
3.   Annual weeds with acquired resistance. In 1981 the first case of resistance to triazine herbicides was reported in this country (Putwain 1982). <u>Senecio vulgaris</u> (groundsel) was found to be resistant to five times the usual dose of the herbicides, this strain of the weed having arisen on a nursery stock holding, as it had in the USA ten years before. This is thought to be a result of the inability to top up herbicides once the crop comes into leaf, and the consequent sub-lethal levels of herbi-cide in the soil resulting in a building up of tolerance to that chemical. In soft fruit crops, and orchards, where contact herbicides can be used, this build up of resistant strain is perhaps of lesser importance, and there are also alternative pre-emergence herbicides which will control the triazine resistant strain of <u>S. vulgaris</u>, e.g. napropamide. Other weed species are also now known to be developing resistance to herbicides, and a strain of <u>Poa annua</u> (annual meadow-grass) has been seen resistant to paraquat and <u>Chamomila suaveoleus</u> (rayless mayweed) to simazine.
4.   Annual weeds - volunteer crops. This subject has already been covered by another paper in this session. It is appropriate to repeat that within an arable situation, with farmers continually searching for alter-native break crops, the possibility of new crop volunteers must always be considered.
5.   Perennial weeds - seed borne. Several of the <u>Epilobium</u> species, most particularly <u>E. ciliatum</u> (American willowherb) which is usually resistant to simazine have increased over the past few years. Screening trials on this genus have been reported (Bailey and Hoogland, 1984) with bromacil and diuron being most effective, though recent Agricultural Development and Advisory Service trials also suggest that oxadiazon, lenacil, and oxyfluorfen can also give excellent control. <u>Malva sylvestris</u> (common mallow) and <u>Hypericum perforatum</u> (perforated St John's wort), both late spring germinators are also being increasingly reported by soft fruit growers. Both are normally controlled pre-emergence by the standard herbicides used in soft fruit, so control can be anticipated if the level of herbicide in the soil can be maintained throughout the germination period.

The most common seed borne perennial is Urtica dioica (stinging nettle), but this is well controlled by the standard residual herbicides pre-emergence of the weed. The established weed is well controlled by triclopyr, but there is considerable danger of crop damage from this volatile herbicide. Control of soft fruit crops is best obtained therefore by spot application of dichlobenil in bush or cane fruit, and hand digging or spot application of mecoprop in strawberries, accepting some crop damage.

6. Stoloniferous perennials. Some of these weeds are susceptible to winter application of paraquat, which can be used in bush and cane fruit, but not overall in strawberries. In this case the only possibility seems to be 2,4-D amine post-harvest at 'growers risk' as there is no label recommendation for this use. Stoloniferous grass weeds are now less of a problem with the availability of specific graminicides, propyzamide and where appropriate, glyphosate.

7. Deep rooted perennials. Weeds such as Convolvulus arvensis and Equisetum arvense (field horsetail) are extremely difficult to eradicate totally and every effort should be made pre-planting to clean the land of such species. Within crops, containment of the weeds is the most practical answer, with the use of such materials as dichlobenil. Lateral spread of the underground organs is likely to continue, even when top growth is controlled using such herbicides or mechanical means, and once treatment is stopped, growth of the weeds will quickly recur (Bailey and Davison, a & b 1984). E. arvense is best controlled pre-planting of soft fruit using aminotriazole applied to a full canopy of soft fronds in early summer. In bush and cane fruit it can best be contained by dichlobenil, and in strawberries propyzamide can give a degree of suppression.

8. Woody perennials. Bush and tree species are occasionally seen in soft fruit plantations. Pre-emergence they are all well controlled by the standard residual herbicide but post-emergence control can only safely be achieved by careful hand work.

1.3 Weed control is rather like mountain climbing; no matter how many times the problem is overcome, or the summit climbed, the mountain still remains for others to climb. With the continual changes in agricultural practices and methods of growing soft fruit, it seems very unlikely that we will ever be in a position to completely overcome some problem weeds.

REFERENCES

1. BAILEY, J.A. and DAVISON, J.G. (1984a) The effect of repeated defoliation for varying periods in four successive years on the shoot growth and underground stems and roots of Convolvulus arvensis. Aspects of Applied Biology, 8, Weed control in fruit crops. pp 9-16.
2. BAILEY, J.A. and DAVISON, J.G. (1984b) The effect of 4 annual applications of chlorthiamid and 2,4-D amine on shoot and root underground stems of Convolvulus arvensis. Aspects of Applied Biology, 8, Weed control in fruit crops pp 17-24.
3. BAILEY, J.A. and HOOGLAND, D. (1984) The response of Epilobium spp. to a range of soil and foliar applied herbicides. Aspects of Applied Biology, 8, Weed control in fruit crops pp 43-52
4. FSPA (1982) Farm Shop and Pick Your Own Association, Which Magazine
5. JONES, A.G. (1984) Current and future weed problems in fruit crops. Aspects of Applied Biology, 8, Weed control in fruit crops pp 1-7
6. PUTWAIN, P.D. (1982) Herbicide resistance in weeds - an inevitable consequence of herbicide use? Proceeding British Crop Protection Council Conference - Weeds, 717-728

Biology and control of *Epilobium ciliatum* Rafin. (Syn.: *E. adenocaulon* Hausskn.)*

R.Bulcke & F.Verstraete
Laboratorium voor Landbouwplantenteelt en Herbologie, Faculteit van de Landbouwwetenschappen, Rijksuniversiteit, Gent, Belgium

M.Van Himme & J.Stryckers
Centrum voor Onkruidonderzoek (I.W.O.N.L.), Faculteit van de Landbouwwetenschappen, Rijksuniversiteit, Gent, Belgium

Summary

The American willowherb, *Epilobium ciliatum* Rafin (syn.: *E. adenocaulon* Hausskn.), is rapidly spreading in Belgium and may occur as a weed in tree nurseries, fruit plantations, maize and at sites with total weed control. This species also occurs in long-term herbicide experiments in fruit plantations where 2-chloro-triazines, alone or in combination with paraquat or aminotriazole, have been applied each year since more than 20 years. These cases concern triazine-resistant biotypes of this species.
In whole plant laboratory bioassay studies a triazine-resistant biotype was about 30-40 times less susceptible to 2-chloro-triazines than a susceptible one. Metribuzine, lenacil and certain urea herbicides were also clearly less active against the triazine-resistant biotype; some other ureas and cell division inhibitors however displayed about equal activity against resistant and susceptible origins.
A paraquat-tolerant biotype was supporting higher rates of paraquat applied post-emergence and higher concentrations during germination. A comparison of 12 biotypes (resistant and susceptible) revealed a high variability between biotypes concerning their leaf colour and dimensions, plant height and fresh matter production although these differences could not always be related to the reaction to s-triazines. Seeds of neither biotype showed any dormancy.

1. Introduction

Willowherbs, *Epilobium* spp., are an increasing problem in perennial crops in the UK (1) and in vines in France (6,11). As their importance as weeds is relatively recent, the information on their response to herbicides is limited to a few reports (1,11,13). Due to the difficulty in distinguishing between the different species, some of the published information deals with unspecified willowherb.

At present the American willowherb, *Epilobium ciliatum* Rafin (syn.: *E. adenocaulon* Hausskn.) is rapidly extending in Belgium. This species is occurring in places with no or a narrow crop rotation and where 2-chloro-s-triazines are regularly applied; this is the case in tree nurseries, long-term herbicide experiments in fruit plantations, maize and at sites with total weed control.

* Research with financial support by the Institute for Encouragement of Scientific Research in Industry and Agriculture (I.W.O.N.L./I.R.S.I.A.)

Studies on the biology of *E. ciliatum* and on its response to herbicides have been published before (1,8,9,13).

The present study deals with: (a) the evolution of *E. ciliatum* populations in a long-term herbicide experiment in a fruit plantation; (b) the reaction of *E. ciliatum* biotypes to soil applied herbicides and to paraquat; (c) a comparison between triazine-resistant and -susceptible biotypes with respect to their germination behaviour, growth and development.

2. Materials and methods

2.1. Origin of *E. ciliatum* biotypes

A brief description of the 12 *E. ciliatum* biotypes mentioned in this study is given in Table I. A collection of these biotypes was set up in the garden of the Faculty of Agricultural Sciences at Gent on April 30, 1984 in pots (Ø 22 cm) filled with a sandy loam soil with the following characteristics: 13.4% clay (0-2 μm), 61.1% loam (2-20 μm), 25.5% sand (>60 μm); 2.16% organic matter (%Cx2); pH(H$_2$O) 5.80, pH(KCl) 4.87 and CEC 7.83 me./100 g. The plants were thinned to 3 per pot; there were 8 replicates.

Table I: Origin of *Epilobium ciliatum* biotypes

Biotype number (x)	Site	Brief description of habitat
1R	Melle 1	Orchard: simazine (4,000 g/ha) since 1958
2R	Melle 2	Orchard: atrazine (4,000 g/ha) since 1961
3R	Meerdonk	Orchard: simazine (2,000 g/ha) since 1962
4R	Melle 3	Roadside: total weed control with simazine
5R	Waarschoot	Tree nursery: simazine (750-1000 g/ha)
6R	Vrasene	Maize: atrazine
7R	Meulebeke 1	Salsify field
8S	Gijzenzele	Ruderal site, on debris of stones
9S	Tielt	Fallow land
10S	Gent	Vegetable garden
11S	Meulebeke 2	Ditch side
12ST	Melle 4	Orchard: paraquat (4 x 1,000 g/ha) since 1971

(x) R = triazine-resistant; S = triazine-susceptible;
    T = paraquat-tolerant

2.2. Reaction to herbicides

2.2.1. Evolution of *E. ciliatum* in a long-term herbicide experiment

The evolution of *E. ciliatum* in a long-term herbicide experiment (Experiment MT 58.6/A), initiated in 1958 at Melle in a bush apple plantation, was studied by counting the number of plants in 32 quadrats (25 cm x 25 cm) per strip. Herbicide applications (see Table II) are carried out yearly in the first half of March. Full details on this experiment have been reported elsewhere (13).

2.2.2. Reaction to soil-applied herbicides

The response of biotypes "1R" (triazine-resistant; see Table I) and "8S" (triazine-susceptible) to soil applied herbicides was studied using a modified small pot bioassay method described elsewhere (10). The herbicide suspensions were mixed uniformly through the air dry sandy loam soil described above; in the untreated control the soil was mixed with an equivalent amount of water. The soil samples were stored overnight in a refrigerator or for a longer period in a deep-freeze. Of either biotype 50 seeds were sown very shallowly (0-1 mm deep) in small plastic pots ($\emptyset$ 6.5 cm). Water was provided by sub-irrigation when necessary. The experimental design was a split-block with 4 replicates. The bioassay experiments were carried out in the greenhouse, where supplementary light was provided for 16 h a day by metal halide lamps (Philips HPI/T/400 W) resulting in a total intensity of about 250 $\mu E.m^{-2}.s^{-1}$. After a growth period of 4 weeks the fresh weight of the aerial parts per pot was determined and, in experiments with photosynthesis inhibitors, divided by the total number of plants emerged to give the fresh weight per plant emerged; for cell division inhibitors the fresh weight per pot was determined. To quantify the differences in response to s-triazines between the susceptible and the resistant biotype, the ED_{50}-values for fresh matter production of aerial parts were determined (Table III) from the regression lines: $y = a + b\log x$ (y = relative yield; x = concentration in w/w). The concentration range used was function of the herbicide and the biotype. In experiments with soil applied a function of the herbicide and the a geometrical series of concentrations per herbicide (Figure 1) was used.

2.2.3. Reaction to paraquat

The response of biotypes "11S" (paraquat-susceptible) and "12ST" (paraquat-tolerant) to paraquat was studied both in a pot experiment and in a germination experiment. The pot experiment, a split-block design with 4 replicates, was initiated on September 4, 1984. Following emergence the plants were thinned to 15 per pot. In addition to an untreated control there were 2 paraquat rates: 300 g and 900 g/ha. Treatments were carried out with a knapsack sprayer (1000 l/ha water; 300 kPa) on October 26, 1984 when the plants had 10-12 leaves and were 12-14 cm high. On November 7, 1984 fresh and dry weight of all aerial parts were taken per pot. The germination experiment with 4 replicates was carried out in petri dishes, each containing 50 seeds, on Schleicher & Schüll nr. 597 and nr. 2282 filter papers. In addition to an untreated control 3 paraquat concentrations were tested: 0.2 ppm, 2 ppm and 20 ppm. After 2 weeks the number of green, chlorotic and dead seedlings and of ungerminated seeds were determined. For statistical analysis the data (%) were transformed to arcsin$\sqrt{\%}$.

2.3. Germination and growth of susceptible and resistant
E. ciliatum biotypes

A germination experiment with seeds of the 12 biotypes, harvested in the collection (see Table I) on August 6, 1984, was set up one day later. Methods and statistical analysis were similar to what is described under 2.2.3. The temperature fluctuated   around 25°C. The % of germinated seeds was determined at regular intervals up to 34 days after setting up the experiment.

For each biotype of the collection (Table I), the mean plant height was determined on August 9, 1984 by measuring 20 plants; on August 20, 1984 the fresh weight of aerial parts was determined on 3 pots per biotype. Leaf characteristics (length, width, area) were assessed for 4 biotypes ("1S", "8S", "11S" and "12S": see Table I) on October 30, 1984 on 25 leaves per biotype taken from the central part of the stem of plants grown up in the greenhouse (sowing date: September 4, 1984). Leaf area was measured with an automatic leaf area meter (type Li-3050A/3). For biotypes "1R" and "8S" the mean number of fruits per plant was determined on August 20, 1984 on material taken from the biotype collection.

Table II. Evolution of the density of *Epilobium ciliatum* populations in a long-term herbicide experiment in a bush apple plantation at Melle (Experiment MT 58.6/A)

Strip nr.	Treatment herbicide	rate (g/ha)	Number of *E. ciliatum* plants per m² in											
			1980		1981		1982		1983		1984		
			9.06	12.09	4.06	16.07	3.06	2.08	26.05	28.07	8.03	25.05	31.07
1.	simazine	4,000	0	10	50	45	90	78	461	229	119	181	45
2.	simazine + paraquat	2,000 + 1,000	0	0	8	7	88	42	460	209	75	47	31
3.	simazine + paraquat	1,000 + 1,000	0	0	1	1	22	13	124	42	17	7	3
4.	simazine + "amitrole-T"(x)	2,000 + 2,000	0	0	24	27	39	53	575	325	94	103	60
5.	atrazine	4,000	0	2	34	67	33	58	356	190	89	80	81
6.	atrazine + paraquat	2,000 + 1,000	0	1	41	37	46	61	360	260	89	106	81
7.	atrazine + "amitrole-T"	2,000 + 2,000	0	14	48	68	48	41	492	210	138	138	94
8.	diuron + paraquat	2,000 + 1,000	0	0	0	0	0	0	0	0	0	0	0
9.	diuron + "amitrole-T"	2,000 + 2,000	0	0	0	0	0	0	0	0	0	0	0
10.	"amitrole-T"	10,000	0	0	0	0	0	0	0	0	0	0	0
11.	untreated (mowing 3-4 times per growing season)		0	0	0	0	0	0	0	0	0	0	0

(x) "amitrole-T" = amitrole + ammonium thiocyanate

3. Results and discussion

3.1. Reaction to herbicides

3.1.1. Evolution of *E. ciliatum* in a long-term herbicide experiment

E. ciliatum was first observed at the end of summer 1980 on some triazine treated strips (Table II) and there has been a very sharp increase in the density of the populations until 1983. The lower number of plants on the simazine + paraquat (1,000 g + 1,000 g/ha) treated strip is due to the high density of other, mainly perennial, plant species. In 1984 a sharp decline in the density of the willowherb stands on strips 1-7 has been observed (Table II); this decline may have several causes like a mowing treatment in June 1983, unfavourable germination conditions in the spring of 1984 and the invasion of strips 1-7 by triazine-resistant annual meadow-grass, *Poa annua* L., since 1983 (12,13). Not a single *E. ciliatum* plant has so far been observed on strips 8-11 (Table II). Strip 11 is mown 3-4 times per growing season and is covered with a dense vegetation of grasses and perennial bread-leaved species, which prevents *E. ciliatum* to invade it. On strips 8 and 9 diuron, a very potent herbicide against *E. ciliatum* (see further), is applied.

3.1.2. Reaction to soil applied herbicides

Biotype "1R" exhibited a high degree of resistance to all 3 triazines tested, as is shown by the values of the resistance factor (Table III). These values of the resistance factor are relatively low if compared with those obtained, with the same bioassay method, for other weed species like e.g. *Poa annua* (12). These whole plant bioassays need confirmation by studies with isolated chloroplasts to find out the existance and degree of resistance at the chloroplast level. So far resistance to triazines has been reported only for the square-stalked willowherb, *Epilobium tetragonum* L. (4,11), although in one study (1) simazine (1,000 g/ha) failed to control *E. ciliatum*.

Table III: ED_{50}-values of 3 s-triazines for fresh matter production of susceptible (=S) and resistant (=R) *Epilobium ciliatum* (1 ppm = 1 mg a.i./kg air dry soil)

Herbicide	ED_{50} (ppm) for biotype		$\dfrac{ED_{50}\ (R)}{ED_{50}\ (S)}$
	1R(x)	8S(x)	
simazine	3.44	0.09	36
atrazine	0.77	0.02	38
cyanazine	2.60	0.09	29

(x) For description of biotypes: see Table I

Metribuzin (a triazinone), hexazinone (a triazine-dione) and lenacil (a uracil) were highly active against the susceptible biotype but much less against the resistant origin (Figure 1). Especially lenacil was well tolerated by the triazine-resistant *E. ciliatum* biotype. In other weed species a similar differential response of resistant and susceptible material to these herbicides has been reported before (2,3).

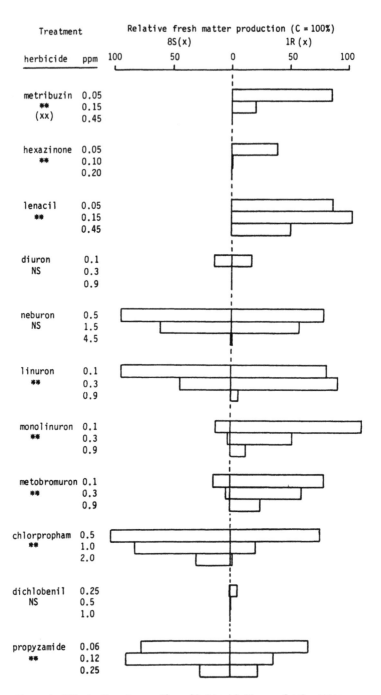

Figure 1: Effect of various soil applied herbicides on fresh matter
production of susceptible (=S) and resistant (=R) *Epilobium ciliatum*

(x) For description of biotypes: see Table I
(xx) For a given herbicide the interaction biotype x herbicide concentration is not significant (NS) of significant at P 0.01 (**).

The response of both biotypes to various urea herbicides was highly variable (Figure 1). Even at low concentrations of diuron, monolinuron and metobromuron the susceptible biotype was nearly completely killed. Linuron, monolinuron and metobromuron were significantly less active on the resistant willowherb whereas diuron and neburon had a comparable activity against the 2 biotypes (Figure 1). In other weed species resistant and susceptible biotypes behave similarly to these urea herbicides (2,3). The results obtained in these bioassay experiments with diuron and neburon confirm the good results obtained on resistant *E. ciliatum* in field and pot experiments (1,13).

The results obtained with herbicides acting primarily on cell division (chlorpropham, dichlobenil and propyzamide) show the very high degree of activity of dichlobenil on both biotypes (Figure 1). The resistant biotype was significantly more susceptible to chlorpropham and to propyzamide (Figure 1), although *E. ciliatum* cannot be classified as being very susceptible to these herbicides as is confirmed by the results of field experiments (13). Published information on experiments carried out in pots as well as in the field shows excellent control achieved with dichlobenil (1,13).

3.1.3. Reaction to paraquat

Although the lower dose of paraquat (300 g/ha) caused distinct phytotoxicity symptoms (necrosis) on biotype "11S" (paraquat-susceptible), its fresh and dry matter production were not inhibited significantly more than was the case for the tolerant biotype "12ST" (Table IV). At the higher dose (800 g/ha) both fresh and dry matter production of the susceptible biotype were significantly inhibited whereas in the resistant biotype, there was no significant further inhibition (Table IV).

Table IV: Effect of post-emergence application of paraquat on a susceptible (=S) and a tolerant (=ST) biotype of *Epilobium ciliatum*

Paraquat rate (g/ha)	Relative production (C = 100%)			
	fresh matter		dry matter	
	biotype 11S (x)	biotype 12ST (x)	biotype 11S	biotype 12ST
0 (=C)	100 a(xx)	100 a	100 a	100 a
300	87 a	77 a	81 a	86 a
900	34 b	73 a	53 b	81 a

(x) For description of biotypes: see Table I
(xx) Means in a column followed by the same letter are not significantly different at the 5% level of probability as determined by Duncan's multiple range test.

Analysis of the results of the germination experiment with paraquat revealed a very significant interaction between biotype and paraquat concentration. Comparison of the percentages of seedlings with green cotyledons clearly shows a very drastic decrease caused by 0.2 ppm paraquat on biotype "11S" (susceptible) whereas at 2.0 ppm still 94.5% of the seedlings of biotype "12ST" (tolerant) had green cotyledons (Table V). Although the physiological basis of variation between the 2 biotypes in susceptibility to paraquat has not yet been investigated, the fact that tolerance is appearing already during germination as well as in later growth stages following a foliar spray, supports the hypothesis that a similar mechanism of tolerance as observed in *Lolium perenne* L. (5) might be involved.

Table V: Effect of paraquat on germination of a susceptible (=S) and a tolerant (=ST) biotype of *Epilobium ciliatum*

Paraquat concentration (ppm)	Mean %							
	seedlings with				dead seedlings		ungerminated seeds	
	green cotyledons		chlorotic cotyledons					
	11S (x)	12ST (x)	11S	12ST	11S	12ST	11S	12ST
0.0 (=C)	98.5 a(xx)	100.0 a	0.0 b	0.0 c	0.0 b	0.0 b	1.5 c	0.0 b
0.2	26.0 b	98.0 b	69.0 a	0.0 c	8.5 a	0.0 b	2.5 c	2.0 b
2.0	17.5 bc	94.5 c	41.5 a	3.0 b	19.0 a	0.0 b	22.0 b	2.5 b
20.0	0.0 c	4.5 d	1.5 b	57.0 a	1.0 b	18.5 a	98.0 a	20.0 a

(x) For description of biotypes: see Table I
(xx) Means in column followed by the same letter are not significantly different at the 5% level of probability as determined by Duncan's multiple range test.

3.2. Germination and growth of susceptible and resistant *E. ciliatum* biotypes

The results of the germination experiment at 34 days from the start, show a nearly complete or complete germination for all 12 biotypes studied (Table VI). *E. ciliatum* has been reported to have no primary dormancy (8,9) but its success as a weed lies, to a great extent, in an efficient secondary dormancy mechanism. Further germination tests under a range of environmental conditions are necessary to permit any conclusions.

Visual observations on leaf colour and growth habit of the plants in the biotype collection (see Table I) revealed that biotypes could be classified in 2 groups. The resistant biotypes "1R" to "7R" and the susceptible "8S" had the same normal green leaf colour, wider leaves and were growing lower (see below); biotype "8S" was clearly growing more vigorously (see below). The triazine-susceptible biotypes "9S" to "11S" and the paraquat-

Table VI: Germination, plant height and fresh matter production of 12 *Epilobium ciliatum* biotypes

Biotype (x)	Germination (%) after 34 days (xx)	Mean plant height (cm) (xx)	Fresh matter (g/pot) (xx)
1R Melle 1	98.5 a	25.6 ab	33.7 ab
2R Melle 2	98.5 a	27.4 bc	29.3 ab
3R Meerdonk	99.0 a	26.9 abc	28.8 a
4R Melle 3	97.5 a	27.3 bc	39.5 ab
5R Waarschoot	99.0 a	26.3 abc	33.6 ab
6R Vrasene	97.0 a	24.3 a	31.1 ab
7R Meulebeke 1	98.0 a	28.9 cd	27.4 a
8S Gijzenzele	99.5 a	31.2 d	57.4 c
9S Tielt	100.0 a	40.2 e	43.5 b
10S Gent	98.5 a	40.3 e	43.3 b
11S Meulebeke 2	98.0 a	41.6 e	34.5 ab
12ST Melle 4	98.5 a	31.5 d	26.2 a

(x) R = triazine-resistant; S = triazine-susceptible; T = paraquat-tolerant
(xx) Means followed by the same letter are not significantly different at the 5% level of probability as determined by Duncan's multiple range test.

tolerant "12ST" had a greygreen leaf colour, narrower leaves and were generally growing higher (see below). *E. ciliatum* has been described as a species with "considerable phenotypic flexibility" (8,9) but in the present
study all biotypes have been growing under the same conditions. So it would
look worthwile to investigate whether or not there is a basis for subdivi-
sion of this *Epilobium* species as has been suggested before (7).

Leaf length, width and area of biotype "8S" (triazine-susceptible) were
significantly higher than those noted for the comparable biotype "1R" (triazine-resistant)(Table VII). The leaves of biotype "11S" and "12ST" (both
with greygreen leaves) were significantly narrower than those of biotypes
"8S" and "1R" (Table VII) which confirms the visual observations mentioned
above.

Table VII: Leaf characteristics of 4 *Epilobium ciliatum* biotypes

Biotype (x)	Mean leaf		
	length (cm)	width (cm)	area (cm^2)
8S Gijzenzele	7.1 a(xx)	2.3 a	11.06 a
1R Melle 1	6.3 b	2.1 b	9.64 b
11S Meulebeke 2	6.5 b	2.0 c	9.48 b
12ST Melle 4	6.4 b	1.8 d	9.30 b

(x) R = triazine-resistant; S = triazine-susceptible;
    T = paraquat-tolerant
(xx) Means in a column followed by the same letter are not signi-
    ficantly different at the 5% level of probability as deter-
    mined by Duncan's multiple range test.

The mean plant height of biotype "8S" to "12ST" (all triazine-susceptible) was significantly higher than that of the triazine-resistant biotypes
"1R" to "7R" (Table VI). The fresh matter production of susceptible biotype
"8S" was significantly higher than that of all triazine-resistant and of all
other susceptible biotypes (Table VI). A comparison of biotype "8S" with
"1R" in a pot experiment carried out in the greenhouse, revealed a much faster initial growth for the susceptible origin (14). Unfortunately only 1
triazine-susceptible biotype ("8S") had the same growth habit like that of all
7 triazine-resistant biotypes ("1R" to "7R").

Biotype "8S" (susceptible) produced significantly more fruits per
plant than the triazine-resistant "1R" (Table VIII).

Table VIII: Mean number of fruits of 2 *Epilobium ciliatum* biotypes

Biotype (x)	Mean number of fruits per plant
1R Melle 1	149 a (xx)
8S Gijzenzele	218 b

(x) R = triazine-resistant; S = triazine-susceptible
(xx) Means followed by the same letter are not significantly
    different at the 5% level of probability as determined
    by Duncan's multiple range test.

4. Conclusions

The results of the present study prove and further confirm that *E. ciliatum* is a very variable species.

E. ciliatum originating from places with a background of 2-chloro-triazine applications was in all cases resistant to the herbicides belonging to this group; prior to the appearance of resistance this species had not been observed at these sites. The differential response of a triazine-resistant and a -susceptible biotype to a range of herbicides belonging to different chemical groups deserves further attention as some of them are commonly used in horticultural crops. In addition to the foregoing, the selection of a paraquat-tolerant biotype following repeated applications of this contact herbicide shows the high potential of this species to appear in a range of man-made habitats.

One triazine-susceptible biotype had a clearly higher "fitness" (faster initial growth, greater height, more biomass, higher seed production) than 7 triazine-resistant biotypes. These results should however be interpreted very cautiously as only 1 susceptible biotype was comparable with the resistant material and as it was originating from a totally different habitat.

REFERENCES

1. BAILEY J.A., RICHARDSON W.G. and JONES A.G. (1982). The response of willowherbs (*Epilobium adenocaulon* and *E. obscurum*) to pre- or post-emergence herbicides. Proceedings 1982 British Crop Protection Conference-Weeds, Volume 1, 299-305.
2. BULCKE R., DE PRAETER H., VAN HIMME M. and STRYCKERS J. (1984). Resistance of annual meadow-grass, *Poa annua* L., to 2-chloro-1,3,5-triazines. Mededelingen van de Faculteit van de Landbouwwetenschappen R.U.G., 49 (3b), 1041-1050.
3. BULCKE R., DE VLEESCHAUWER J., VERCRUYSSE J. and STRYCKERS J. (1985). Comparison between triazine-resistant and -susceptible biotypes of *Chenopodium album* L. and *Solanum nigrum* L. Mededelingen van de Faculteit van de Landbouwwetenschappen R.U.G., 50, (in print).
4. GASQUEZ J. (1981). Evolution de la résistance aux triazines chez les espèces annuelles: la résistance chloroplastique. Compte rendu de la 11ème Conférence du Columa, Tome IV, 980-1008.
5. HARVEY B.M.R. and HARPER D.B. (1982). Tolerance to bipyridilium herbicides. In: "Herbicide resistance in plants", LeBARON H.M. & GRESSEL J. (editors), New York: John Wiley & Sons, 215-233.
6. HEINZLE Y. (1981). Evolution de la flore dans le vignoble française. Résultats d'une enquête. Compte rendu de la 11ème Conférence du Columa, Tome III, 697-705.
7. HEUKELS H. en VAN OOSTSTROOM S.J. (1977). Flora van Nederland. Groningen: Wolters-Noordhoff, 19e druk, 925 p.
8. MYERSCOUGH P.J. and WHITEHEAD F.H. (1966). Comparative biology of *Tussilago farfara* L., *Chamaenerion angustifolium* (L.)Scop., *Epilobium montanum* L. and *E. adenocaulon* Hausskn. I. General biology and germination. New Phytologist, 65, 192-210.
9. MYERSCOUGH P.J. and WHITEHEAD F.H. (1967). Comparative biology of *Tussilago farfara* L., *Chamaenerion angustifolium* (L.) Scop., *Epilobium montanum* L. and *E. adenocaulon* Hausskn. II. Growth and ecology. New Phytologist, 66, 785-823.
10. NYFFELER A., GERBER H., HURLE K., PESTEMER W. and SCHMIDT R.R. (1982). Collaborative studies of dose-response curves obtained with different bio-assay methods for soil-applied herbicides. Weed Research, 22(4), 213-222.

11. ROUAS G. (1981). Essai de lutte contre l'épilobe tétragone (*Epilobium tetragonum*) et la morelle noire (*Solanum nigrum*) en Champagne. Compte rendu de la 11ème Conférence du Columa, Tome III, 635-641.
12. VAN ACKER E. (1985). Triazine-resistent straatgras: biologie, bestrijding en betekenis van de ED$_{50}$-waarde en resistentiefactor. Landbouwhogeschool Wageningen, Vakgroep Vegetatiekunde, Plantenoecologie en Onkruidkunde, 74 p.
13. VAN HIMME M., STRYCKERS J. en BULCKE R. (1983 en 1985). Bespreking van de resultaten bereikt door het Centrum voor Onkruidkunde tijdens het teeltjaar 1981-1982 en 1983-1984. R.U.G. Fakulteit van de Landbouwwetenschappen, Centrum voor Onkruidonderzoek. Mededelingen nrs. 38 en 42, 206 p. en 160 p.
14. VERSTRAETE F. (1985). Biologie en bestrijding van gewimperde basterdwederik, *Epilobium ciliatum* Rafin. Afstudeerwerk Academiejaar 1984-1985, Faculteit van de Landbouwwetenschappen R.U.G., 122 p.

Session 3
Weed control in strawberries

Chairman: G.Noye

Recent developments in weed control in soft fruits in the Netherlands

H.Naber

Plant Protection Service, Centrum voor Agribiologisch Onderzoek, Wageningen, Netherlands

Summary

During many years the methodology of chemical weed control in soft fruit crops was rather stable, being satisfactory from the point of view of efficacy and economy.
Weed control in strawberry was based on the three herbicides lenacil, simazine and phenmedipham and in woody soft fruit crops on simazine, paraquat and amitrole.
Due to changes in crop husbrandry, the introduction of new varieties and the continuous use of the same herbicides new problems arose, as varietal susceptibility to herbicides, persistence in soil, resistance of certain weeds and new weed species invading.
The need for new and more herbicides became evident and at the same time it was felt that the old ones could not be missed. Screening activities were started with herbicides already approved for applications in other crops to find additional chemical solutions.
The new graminicides were tested and they showed a high level of selectivity and efficacy. Research was also started to look for alternatives for the persistent herbicide paraquat. This report deals with recent developments in chemical weed control in soft fruit in the Netherlands and discusses the results of field experiments.

1.Introduction

In the Netherlands soft fruit cultures belong to the minor crops and vine growing is negligible.
In table I is shown that acreages of soft fruit decreased from 8570 ha in 1950 to 2470 ha in 1984. Strawberry is the most important of these crops with more than 75% of the area.
The interest of agrochemical industries for minor crops is restricted because they offer only a small market for pesticides. Nevertheless there was and is a good coöperation between governmental institutes, growers organisations and chemical firms in finding suitable pesticides for soft fruit and other minor crops. As a rule only those pesticides are taken into research for usage in minor crops which are already approved for other crops.
Herbicides have been used for weed control in all soft fruit crops for many years. Little in the methodology of chemical weed control has been changed since these substances were introduced in practice. During the last five years new problems have arisen due to changes in growing techniques, the introduction of new crop varieties, the occurrence of new weeds, and problems arising from continuous herbicide usage. In this paper a review is given of the present situation in chemical weed control in the different soft fruit crops in the Netherlands and results are presented of some screening trials and experimental work in finding solutions for the weed problems.

2. Weed control practices in woody soft fruit

In the different woody soft fruit crops the methodology of chemical
weed control is more or less the same. Simazine, paraquat and amitrole
are the three herbicides on which the system is based.
In spring the residual herbicide simazine is applied at a rate of 0.5 -
1.5 kg/ha to prevent growth of annual weeds. When weed vegetation is already
present at that moment the contact herbicide paraquat is added at an
amount of 0.6 - 1.0 kg/ha, or this herbicide is used later on during
spring or summer.
In autumn the systemic acting leaf herbicide amitrole is used, especially
against perennial weeds at an amount of 3 kg/ha. The time of application
of amitrole is restricted from post-harvest until November 1th, due to
the fact that amitrole is approved on a no-residue level.
When paraquat or amitrole are applied between and under bushes and
trunks of woody soft fruit species a protecting screen is necessary to
prevent contact of the crop with the spray droplets.
Not only the leaves but also the buds, especially the green buds of
black currants can take up amitrole. Red currants are more and more
grown in hedges instead of bushes. In this husbandry system paraquat is
not only used for weed control but also for the control of suckers. Also
gooseberry is grown in hedges and growers should like to control
suckers too with paraquat, but no experimental information is available.
In brambles and raspberry which traditionally have been grown in hedges
sucker control is done mechanically before mid May.

Other herbicides than the main-three already mentioned are used such as
the residual herbicide diuron at a rate of 0.8 - 1.6 kg/ha. Diuron has
a lower selectivity than simazine in woody soft fruit and is even too
dangerous for blackberry. Contact with green parts of the crops must be
prevented, otherwise damage occurs.
It can be predicted that diuron will become more important for the
control of Poa annua because the incidence of resistance of this weed
to simazine is increasing. Also for the control of Epilobium parviflorum
and Amaranthus lividus, diuron is the only herbicide that is suitable, but
it has to be applied in combination with simazine, otherwise the escape of
certain weed species will occur. For the grower it means that the costs of
chemical weed control are more than doubled and it is unfavourable also
from the environmental point of view.

Dichlobenil and chlorthiamid have been registered for many years for
use in red and black currants and in gooseberry for the control of
Equisetum arvense, Cirsum arvense and Tussilago farfara. These granular
herbicides are only recommended for spot treatment but are hardly used
in practice.
For the control of Calystegia sepium in red currants MCPA at a rate of
1 kg/ha can be applied after harvest as soon as the endbuds are closed.

The new graminicides alloxydim-sodium, sethoxydim and fluazifop-butyl
are all approved for the control of grasses in woody soft fruits.
They are mainly used against Agropyron repens and in case Echinochloa
crus-galli causes problems, but until now their use is not of
importance. Time of application is restricted to the periods before
flowering and after harvest.

The future of paraquat registration is rather unsure due to the persistence of this herbicide in the soil. Research is necessary to find out whether the new contact-herbicide glufosinate-ammonium can replace paraquat in all its applications.

3. Weed control practices in strawberry

The methodology of chemical weed control in strawberry has not been changed since the introduction of phenmedipham. Very roughly it can be said that a few weeks after planting in August the residual herbicide lenacil is sprayed at a rate of 0.8 - 1.6 kg/ha. When young weeds occur during autumn the contact herbicide phenmedipham is applied at a rate of 0.94 kg/ha.
In the month of March (late winter - early spring) simazine is used in a rate of 0.25 - 0.38 kg/ha. For crops that suffered from frost or water the application of simazine is too risky. When necessary phenmedipham is applied before flowering of the crop especially when Urtica urens is present.
This system of weed control is carried out in the annual strawberry culture for fresh consumption, which now includes 90% of the total area (excluding strawberry under glass and the propagation culture). Besides the big three herbicides, chloroxuron can also be used but only in autumn and mainly against Stellaria media.

Due to the long persistence of lenacil in soil this product can not be applied within half a year before the end of the culture to prevent damage in the following crop. So, in the annual crop it can only be used in summer or autumn short after planting. In the perennial crop also spring application can be carried out. After harvest of the perennial strawberry simazine can be used at a rate of 0.5 - 0.75 kg/ha (the double amount from the annual crop) and when necessary in a tankmix with 0.94 kg/ha phenmedipham. Also lenacil can be used at that moment at a higher rate (1.2 - 2 kg per ha) in comparison to the annual crop.

On the propagation fields the same herbicides are used and the same system is followed as in the annual crop. When the propagation field stays for another year the system of the perennial culture is followed.

On waiting beds lenacil 0.8 kg per ha is applied 1 to 2 weeks after planting, but only when strawberries will be grown in the same field in the following year.   In case of growing this type of strawberry in a rotation, only chloroxuron 3 - 3.5 kg/ha and phenmedipham 0.94 kg/ha can be used.

In the late-summer culture of strawberry (cold stored plants) also lenacil 0.8 kg/ha (so in a low dose, when no following crop will be grown in the same year) and phenmedipham 0.94 kg/ha are used for weed control.

In case of continuous bearing strawberry chloroxuron can not be used and the possibilities for the other herbicides are restricted.

4. Soil fumigation

Due to the fact that strawberry growing is mainly done on the same field every year of (monoculture) problems can arise with certain nematodes and fungi. Therefore soil fumigation with 1.3-dichloropropene or metham-sodium is carried out once in 2 to 3 years. Both fumigants have a good side-effect on weeds, metham-sodium mainly on annual weeds, 1.3-dichloropropene on Agropyron repens.
In practice it means that after soil fumigation between two crops, it is often unnecessary to apply lenacil for weedcontrol.

5. Search for new herbicides in strawberry

As described earlier in this paper the methodology of chemical weed control in strawberry is based on 3 to 4 herbicides, all having one or more disadvantages:
- Lenacil being rather persistent in the soil and showing suscept bility of certain (new) cultivars to this herbicide and therefore only suitable for one treatment in a year at a rather low rate.
- Simazine with a marginal tolerance, thus only suitable at very low rates in a well-established healthy crop. Also resistance to simazine of weeds such as Poa annua, Solanum nigrum and Senecio vulgaris and tolerance of Viola arvensis, Lamium purpureum and Veronica species to this herbicide and lenacil are causing problems in practice.
- Phenmedipham being only useful in combination or in succession with the other two.
- Chloroxuron always causing some foliar damage and no more allowed in production fields during springtime due to the no-residue tolerance. Therefore the need for new and more herbicides became evident.

Already in 1974 cyanazine was tested at a rate of 0.5 and 1.0 kg/ha in the cultivar Confitura. This residual herbicide was applied in spring in comparison with simazine. Both rates of cyanazine caused severe damage to the crop and simazine did not.

From 1977 onwards screening activities started at the Experimental Garden "Noord Brabant" at Breda, the centre of the strawberry growing area.
In table II the results of testing selectivity of 20 herbicides are presented. This field experiment was carried out in the month of May in newly planted strawberry, cultivar Sivetta on a waiting bed on a sandy soil type with 3% organic matter.
In the same year those herbicides that looked promising were tested again in the month of August in a newly planted production field, also of the cultivar Sivetta (table III). The herbicides chloridazon and propazine showed good weed control, but caused rather severe damage. Propachlor, neburon and asulam showed a rather good weed control and light to severe damage.

Therefore the experiment was continued in 1978 with the same herbicides at lower rates and the herbicide cycloate was added. The treatments were applied in August in a new planted production field of the cultivars Confitura en Redgauntlet.(table IV).

Neburon and asulam were again shown to be too phytotoxic to strawberry.
Chloridazon (0.65 - 1.3 kg/ha) and propazine (0.25 - 0.38 kg/ha) caused
hardly any damage to the crop. Weed control of chloridazon was bad and
from propazine good, comparable to lenacil 1.2 kg/ha. The successive
applications of first cycloate 2.88 kg/ha and on the same day
phenmedipham 0.94 kg/ha and another successive application of
chloroxuron 3.0 kg/ha and phenmedipham 0.94 gave good weed control
without phytotoxicity. Due to the disappointing control of weeds with
chloridazon and the fact that propazine was no longer marketed, further
research with these herbicides was stopped.

In 1979 tank mixtures of cycloate/phenmedipham and chloroxuron/
phenmedipham were tested in the month of May in the cultivar Confitura
(table V). Both combinations caused severe phytotoxicity. In another
field experiment on the Experimental Garden at Kesteren (riverclay soil)
in the cultivar Sivetta no phytotoxicity was observed after treatments
with these tank mixtures (table VI).
Also in 1979 in the month of June tank mixtures of phenmedipham 0.94 kg/ha
with chloroxuron 3 kg/ha. respectively with cycloate 2.88 kg/ha, with
lenacil 0.8 kg/ha and with lenacil 0.4 kg/ha were tested in 3 cultivars:
Korona. Bogota en Sivetta (table VII). This field trial was carried out
on the Experimental Garden at Breda in a late crop (cold stored plants)
of strawberry. Lenacil in a tank mix with phenmedipham showed to be safe in
all cultivars. Only light phytotoxity was observed at the highest rate
in Bogota. Cycloate/phenmedipham caused damage in Bogota and Korona and
chloroxuron/phenmedipham in all cultivars. In the cultivar Korona
yield depression was established with chloroxuron, cycloate and the
highest rate of lenacil, all in a tank mix with phenmedipham.

After three years of screening on selectivity and efficacy it was
concluded that none of the tested herbicides could improve the
methodology of chemical weed control in strawberry.

In 1982 new research was started at Breda with split and successive
applications (table VIII) of lenacil and simazine and also mixtures
with propachlor and propham. Also ethofumesate and metamitron were
included in this field experiment. The herbicides were applied in the
middle of September in a newly planted strawberry crop, cultivar
Gorella. The second treatment occured at the beginning of November. In
March 1984 all applications were followed by a third treatment with
0.25 kg/ha simazine. Only propham 2.5 kg/ha (in combination with other
herbicides) caused growth inhibition, the other herbicides and split and
successive applications did not. Ethofumesate 0.4 kg/ha showed weak
weed control and metamitron 1.4 kg/ha good weed control as did all other
applications.

In 1983 the emphasis was laid on further development of metamitron in
rates of 1.4 - 3.5 kg/ha in combination with phenmedipham 0.94 kg/ha and
in comparison with lenacil 1.2 kg/ha combined with phenmedipham.
The cultivar Gorella was planted August 3 and sprayings were carried out
on August 16 (table IX).
All treatments were insufficient from a weed control point of view but
lenacil/phenmedipham was the best one. Metamitron caused phytotoxicity at
the highest rate of 3.5 kg/ha in combination with phenmedipham.

Conclusions: It is remarkable that of all herbicides tested only the sugarbeetherbicides seem to be more or less selective in strawberry. The value of chloridazon and cycloate is restricted and does not need further attention. More promising seem to be metamitron and ethofumesate. Looking at the character of these herbicides, metamitron should be further researched for application in spring in mixture with phenmedipham and ethofumesate for application in late summer or autumn in a newly planted crop alone or in mixture with lenacil.

6. Control of volunteer cereal from straw

Volunteer cereal plants germinated from seeds in straw used in strawberry have to be controlled, otherwise competion will occur and picking will be rather difficult. Straw is not only used to cover the soil in order to keep the fruits clean, but also to retard the strawberry crop resulting in a later harvest at a time of higher prices. During two years the new graminicides were taken into official testing for registration by the Plant Protection Service. The field trials were executed by the local officers in different parts of the country. In table X and table XI the data from these field experiments are presented.

In 1981 the addition of mineral oil (11 E oil) to alloxydim sodium and sethoxydim and a wetting agent (Agral LN) to fluazifop-butyl were tested. It became clear that on one hand mineral oil improved the effect of alloxydim-sodium, because 1.13 kg + oil was as good as 1.5 kg without oil or even better. On the other hand slight symptoms of phytotoxicity occurred existing of reddish spots on the oldest leaves and necrotic leafmargins when oil was added.
The same type of phytotoxicity, but less intensive was observed when sethoxydim was mixed with oil. Also for this graminicide the lower rate of 0.37 kg + oil was as good as 0.46 kg without oil.
Fluazifop-butyl gave the best control of volunteer cereal. Also here it was clear that the lower amount of 0.63 kg + wetting agent was as good and sometimes even better then 0.75 kg without wetting agent. No phytotoxicity was observed, either with the graminicide alone, or when the wetting agent was added.

In 1982 the official testing was continued with alloxydim-sodium at the same rate per ha as in the year before, sethoxydim at a higher rate and fluazifop-butyl at a lower rate ; the two primer herbicides were mixed with a mineral oil and the latter with a wetting agent.
The control of cereal volunteer plants was again best with fluazifop-butyl. The second best was sethoxydim and ,in 4 out 6 trials, a bit weaker than the others was alloxydim-sodium. The same type and rate of phytotoxicity could be observed as in 1981 with alloxydim-sodium. Again with sethoxydim the symptoms were less frequent and weaker and again no phytotoxicity was found after using fluazifop-butyl.

Present state of graminicides: The three graminicides mentioned above are now officially registered for weed control in strawberry not only for the control of cereal volunteer but also for the control of Agropyron repens and annual grasses such as Echinochloa crus-galli.
Alloxydim-sodium and fluazifop-butyl are allowed to be applied before flowering and after harvest. Sethoxydim can also be applied during and after flowering until 3 weeks before harvest.

7. Control of Rorippa silvestris

One of the main perennial weeds in strawberry is Rorippa silvestris.
a species belonging to the family Cruciferae and producing rootstocks.
No herbicides which are selective in strawberry are available for control, so control has to be done after harvest before planting a new
strawberry crop. This weed species is already known for a long time from
flowerbulb cultures. Also in flowerbulbs selective herbicides for this
purpose are not available and therefore chemical control is carried out
in the period between two crops.
Already for more then twenty years a product named Antikiek, being a
mixture of 2.4-D and MCPA has been used for control of this weed. The
product contains 103 g/l 2.4-D and 234 g/l MCPA and is formulated as a
dimethylamine salt. Per ha 13 l is used that means 1.3 kg a.i. of
2,4-D and 3 kg a.i. of MCPA which is a very high dosage. Screening
activities in flowerbulbs during last years with all types of systemic
leaf applied herbicides (glyphosate. amitrole, asulam, hormone types)
proved again that this 2,4-D/MCPA mixture is the best one.

At the Experimental Garden at Breda field trials were started in 1983 to
find out if 2,4-D/MCPA can be used safely before a new crop is planted.
Therefore two rates: the normal and the double were used and three
intervals between spraying and planting: 2, 4 and 6 weeks were included.
From the results presented in table XII can be concluded that an
interval of 2 weeks is too short: both rates caused damage to the new
crop. After 4 weeks the double rate still gave an unacceptable damage
but the normal rate seemed to be safe. After an interval of 6 weeks
both normal and double rate were safe for the newly planted crop.
During summer ,breakdown in the soil of these hormone type of herbicides is
so fast that the phytotoxic residues have disappeared after 4-6 weeks.
In another experiment efficacy testing was carried out with 2,4-D/MCPA,
glyphosate and the soil fumigant metham-sodium. The results are presented
in table XIII.
Again it was proved that 2,4-D/MCPA is very effective against Rorippa
silvestris and that glyphosate had almost no effect at all and the
effect of metham-sodium was also bad.

Present state: In practice 2,4-D/MCPA has to be sprayed directly after
the last picking as an overall spray on the weed and the old crop plants.
After 3 weeks the old crop can be removed and soilcultivation can be
carried out. After 5 weeks planting of a new strawberry crop is possible.

8. Acknowledgments

Thanks are due to Mr. J. de Bruyn, fieldresearchworker of the
Experimental Garden "Noord Brabant" at Breda and Mr. W.J. Alofs, plant
protection officer from the Horticultural Advisory Service at Tilburg
who did most of the field testing and to Mr. D. van Staalduine, former
research worker on weed control in fruit crops and minor crops at the
Centre of Agrobiological Research (CABO), Wageningen for his assistence
in preparing this paper.

Table I: ACREAGES OF SOFT FRUITS (ha)

Crop	1960	1970	1979	1984
Strawberry	4883	2803	1862	1886[1]
Red currant	1574	553	195	240
Black currant	1438	180	33	60
Raspberry	1319	410	72	50
Blackberry	*[2]	*[2]	49	140
Other crops	386	207	172	95
Total	9600	4153	2383	2471

1) Not included 153 ha under glass,
   and 154 ha for propagation

2) * = included in other crops

Table II: SCREENING HERBICIDES IN STRAWBERRY, cv. SIVETTA
(waiting bed treated two weeks after planting; May 1977)

Trade name	Active ingredient	Rate per ha	Phytotoxicity
Venzar	lenacil	1.2 kg	none
Gesatop	simazine	0.5 kg	none
Pyramin	chloridazon	1.95 kg	none
Patoran	metobromuron	1.5 kg	none
Dacthal	chlorothal-dimethyl	7.5 kg	none
Legurame	carbetamide	2.1 kg	none
Treflan	trifluraline	0.96 kg	none
Ramrod	propachlor	4.55 kg	light
Campagard	prometryne/propazine	0.5/0.3 kg	severe
Gesamil	propazine	0.5 kg	severe
Dicuran	chlortoluron	1.6 kg	severe
Tribunil	methabenzthiazuron	2.1 kg	severe
Dosanex	metoxuron	2.4 kg	severe
Asulox	asulam	2.4 kg	severe
Kloben	neburon	4.8 kg	severe
Igran	terbutryne	1.5 kg	crop killed
Igrater	metobromuron/terbutryne	1.0/1.0 kg	crop killed
Arelon	isoproturon	1.5 kg	crop killed
Sencor	metribuzine	0.7 kg	crop killed
Tok E 25	nitrophen	1.7 kg	crop killed

Table III: SCREENING HERBICIDES IN STRAWBERRY, cv. SIVETTA
(production field, treatment two weeks after planting;
August 1977)

Trade name	Active ingredient	Rate per ha	Phytotoxicity	Efficacy
Venzar	lenacil	1.2 kg	none	very good
Pyramin	chloridazon	1.95 kg	severe	very good
Dacthal	chlorthal-dimethyl	7.5 kg	none	very bad
Gesamil	propazine	0.5 kg	severe	very good
Ramrod	propachlor	4.55 kg	light	good, except on grasses
Asulox	asulam	1.6 kg	severe	good, except on Urtica urens
Kloben	neburon	3.6 kg	severe	good, except on grasses

Table IV: TESTING HERBICIDES IN DIFFERENT RATES IN STRAWBERRY AUGUST 1978
(Cv. Confitura and Redgauntlet, treatment two weeks after
planting)

Trade name	Active ingredient	Rate per ha	Weed control	Phytotoxicity
Venzar	lenacil	1.2 kg	good	none
Pyramin	chloridazon	0.65 kg	bad	none
Pyramin	chloridazon	1.3 kg	bad	light
Gesamil	propazine	0.25 kg	good	none
Gesamil	propazine	0.38 kg	good	light
Asulox	asulam	0.6 kg	bad	severe
Asulox	asulam	1.2 kg	bad	severe
Kloben	neburon	1.5 kg	bad	light
Kloben	neburon	3.0 kg	moderate	severe
Ro Neet/ Betanal	cycloate/ phenmedipham	2.88/ 0.9 kg	good	none
Tenoran/ Betanal	chloroxuron/ phenmedipham	2.0/ 0.94 kg	good	none

Table V: TESTING HERBICIDE MIXTURES IN SPRING 1979

(Strawberry cultivar Confitura; Exp. Garden Breda)

Herbicides	Rate per ha	Date of application	Weed control	Phytotoxicity
simazine	0.25 kg	April 2	very good	none
cycloate/ phenmedipham	2.88/ 0.94 kg	May 8	good	severe
chloroxuron/ phenmedipham	3.0/ 0.94 kg	May 8	good	severe

Table VI: TESTING HERBICIDE MIXTURES IN SPRING 1979

(Strawberry cultivar Sivetta; Exp. Garden Kesteren)

Herbicides	Rate per ha	Date of application	Weed control	Phytotoxicity
lenacil	0.8 kg	April 25	good	none
cycloate/ phenmedipham	2.88/ 0.94 kg	April 25	good	none
chloroxuron/ phenmedipham	3.0 0.94 kg	April 25	good	none
simazine	0.25 kg	April 25	good	yellowing
simazine	0.25 kg	March 26	good	none
simazine/ chloroxuron/ phenmedipham	0.1/ 1.3/ 0.94 kg	April 25	good	none

Table VII: TESTING TANK MIXES ON VARIETAL SUSCEPTABILITY

(Late strawberry culture, planted June 6; treatments June 29, Yields in kg/100 m^2 (2 replicates)   1979)

Herbicides	Rates kg/ha	Korona Phytotox.	Yield	Bogota Phytotox.	Yield	Sivetta Phytotox.	Yield
lenacil/ phenmedipham	0.8/ 0.94	none	179	little	140	none	197
lenacil/ phenmedipham	0.4/ 0.94	none	193	none	143	none	187
chloroxuron/ phenmedipham	3.0/ 0.94	severe	133	severe	145	little	181
cycloate/ phenmedipham	2.88/ 0.94	moderate	148	moderate	124	none	178

Table VIII: SPLIT AND SUCCESSIVE APPLICATIONS IN AUTUMN

(Strawberry cultivar Gorella, planted August 4, 1982)

Herbicide Treated Sept. 16	Rate per ha		Herbicide Treated Nov. 8	Rate per ha		Weed control	Yield kg/100 m²
lenacil	0.4	kg	lenacil	0.4	kg	good	276
lenacil	1.2	kg				good	300
lenacil	0.4	kg	simazine	0.25	kg	good	321
lenacil/ propham	0.4/ 2.5	kg	lenacil	0.8	kg	good	277 *)
ethofumesate	0.4	kg				moderate	278
metamitron	1.4	kg				good	302
propachlor/ propham	3.25/ 2.5	kg	lenacil	0.8	kg	good	259 *)
simazine	0.125	kg	simazine	0.25	kg	good	280
simazine/ propachlor	0.125/ 3.25	kg	simazine	0.25	kg	good	282
simazine	0.25	kg				good	313
simazine/ lenacil	0.125/ 0.4	kg				good	326
Treated Aug. 25							
lenacil/ phenmedipham	1.2/ 0.94	kg				good	326
untreated hoeing							285

All plots were sprayed again at March 3, 1983 with 0.25 kg/ha simazine

*) Already in the autumn growth inhibition was observed

Table IX: TESTING METAMITRON ON STRAWBERRY, cv. GORELLA 1983/1984

treated *)/August 16			treated March 30		Crop stand	Weed control
metamitron	1.4 kg	———	metamitron	0.7 kg	good	3
metamitron	2.1 kg	———	metamitron	1.4 kg	good	4
metamitron	3.5 kg	———	metamitron	2.1 kg	damage	4
metamitron	1.4 kg	———	metamitron	2.8 kg	good	5
metamitron	2.1 kg	———	metamitron	3.5 kg	damage	5
lenacil	1.2 kg	———	simazine	0.25 kg	good	8

*) All treatments in combination with 0.94 kg/ha phenmedipham

Table X: CONTROL OF VOLUNTEER CEREAL IN STRAWBERRY

(results of 7 fieldexperiments in 3 replications, 1981)

Herbicide	Rate per ha	Results (10 = 100% control)						
		I	II	III	IV	V	VI	VII
alloxydim	1.50 kg	4,5	9	8.5	9	5.5	4	4
alloxydim/ oil	1.13 kg/ 5 l	8	9	9	9.5	6	4	3.5
sethoxydim	0.46 kg	7	9	7	9.5	7.5	4	1
sethoxydim/ oil	0.37 kg/	8	9	9.5	8	8	4	4
fluazifop-butyl	0.75 kg	9	9	10	10	8	10	9
fluazifop-butyl/ wetting agent	0.63 kg/	9	9	9.5	10	9	10	9

Exp.	Cultivar	Input straw	Species and stage of cereal volunteer	
I	Tago	March 6	wheat	tillering
II	Sivetta	May 5	wheat	2-3 leaves
III	Sivetta	June 6	wheat	3-4 leaves
IV	Bogota	April 15	wheat	4 leaves
V	Tago	Febr. 20	rye	4 leaves
VI	Tago	Jan. 5	wheat	4 leaves
VII	Tago	Jan. 5	wheat	tillering

Exp.	Data of treatment	Weed cover	Stage of the crop
I	May 19	rather high	begin flowering
II	June 6	low	end of flowering
III	July 8	high	1 week before picking
IV	June 2	moderate	end of flowering
V	May 20	moderate	begin flowering
VI	May 8	moderate	before flowering
VII	June 4	very high	end flowering

Table XI: CONTROL OF VOLUNTEER CEREAL IN STRAWBERRY
(results of 6 field experiments in 3 replications; 1982)

Herbicides	Rate per ha	I	II	III	IV	V	VI
alloxydim/ oil	1.13 kg/ 5 l	9	9	10	8	6	4.5
sethoxydim/ oil	0.46 kg/ oil	9.5	9	10	9	8	7
fluazifop-butyl/ wetting agent	0.50 kg/	9.5	9	10	10	9	7

Exp.	Cultivar	Input straw	Species and development of cereal volunteer	
I	Elvira	May 31	wheat	early tillering
II	Sivetta	June 1	wheat	early tillering
III	Tago	April 21	oat	3-4 leaves
IV	Bogota	April 27	wheat	early tillering
V	Tago	Febr. 5	wheat	20 - 25 cm
VI	Tago	Febr. 15	wheat	full tillering

Exp.	Data of treatment	weed cover	Crop stage
I	June 21	1-20%	end flowering
II	June 21	2-10%	end flowering
III	June 2	10-70 pl/m^2	before flowering
IV	June 8	1-20%	end flowering
V	June 21	5-40%	2.5 week before picking
VI	June 4	5-35%	end flowering

Table XII: RESIDUAL EFFECT OF 2,4-D/MCPA ON STRAWBERRY, cv. GORELLA

Rate per ha 2,4-D/MCPA	Interval	Crop stand	Number of runners	Killed plants
1.3/3 kg	2 weeks	4.3	7	3
2.6/6 kg	2 weeks	3.4	8	4
1.3/3 kg	4 weeks	6.1	15	1
2.6/6 kg	4 weeks	3.7	8	2
1.3/3 kg	6 weeks	6.9	19	1
2.6/6 kg	6 weeks	7.1	21	1
untreated		5.3	16	1

Table XIII: CONTROL OF RORIPPA SILVESTRIS

Herbicide/Fumigant	Rate per ha	Number of weeds/plot
metham-sodium	1000 l	50
2.4-D/MCPA	1.3/3 kg	0
glyphosate	2.16 kg	50
metham-sodium + 2.4-D/MCPA	1000 l + 1.3/3 kg	0
metham-sodium + glyphosate	1000 l + 2.16 kg	11
2.4-D/MCPA + glyphosate	1.3/3 kg + 2.16 kg	0
untreated		>50

Developments in weed control due to changing growing methods in soft fruit in the Netherlands

H.A.Th.Van der Scheer
Research Station for Fruit Growing, Wilhelminadorp, Netherlands

Summary

In the period 1965-1980 strawberry growing in The Netherlands changed from a perennial system with plants producing for the processing industry to an annual one with plants producing for the fresh market. Nowadays these plants are grown annually on specialized holdings on which crop rotation is often no longer part of the growing system. An other major change in strawberry growing comprises the prolongation of the picking period, using perforated plastic film as a cover of the plants in spring, using cold stored plants, and using ever bearing varieties. In woody soft fruit the growth of red currants and goose-berries in hedgerows, the upright growth of new canes in thornless blackberries and the introduction of grass strips on bigger holdings being partial to mechanization, are the major changes in the growing systems. The consequences of the changes for the weed control are discussed.

1 Introduction

Herbicides have been used for weed control in all soft fruit crops for many years. It started about 1960 with the introduction of simazine in strawberry growing and of simazine and MCPA in woody soft fruit growing (6). In the preceding period soil tillage was labour intensive and contributed for 20%, respectively 18%, in the unit costs of strawberry and woody soft fruit crops. Nowadays weed control in strawberry is based on three herbicides, viz. lenacil, simazine and phenmedipham, and in woody soft fruit crops on simazine, paraquat and amitrole. The introduction of herbicides did not only reduce the unit costs, but also raised the yield of the woody soft fruit crops, because superficially growing roots are no longer damaged as they are in case of soil tillage.

2 Annual growing of strawberry

In 1960 the acreage of soft fruit crops comprised about 9600 ha. However, since 1965, that acreage decreased strongly. Growing for the processing industry was no longer profitable, due to the import of low--priced fruit pulp and in recent years deep-frozen fruit from East-European countries, particularly Poland, by the processors. To-day the acreage comprises about 2500 ha, of which 2000 ha of strawberries, being the most important soft fruit crop in The Netherlands (7). Nowadays most of the fruit is grown for the fresh market, where large berries of a good quality are preferred. As a consequence the strawberry growing system changed from a perennial into an annual one. In the annual crop the residual herbicide lenacil cannot be applied within half a year before the end of the culture to prevent damage in the following crop. Therefore in spring simazine (0.25-0.38 kg/ha) and phenmedipham (0.94 kg/ha) are used, and in case weeds

85

are present they have to be hand-hoed out of the plant rows and to be
sprayed with paraquat (0.4-0.6 1/ha) using a protecting screen between the
plant rows. However, some of the weed species (Viola arvensis, Rorippa
silvestris and resistant biotypes of Poa annua) are not, or not any more,
controlled by these three herbicides. When they are present, hand-hoeing is
the only solution (3). In runner fields (in 1985 about 200 ha) the presence
of Rorippa silvestris in spring may lead to the decision to omit the
digging up of the strawberry plants and to destroy them in the end (2).

The monoculture of strawberry can cause problems with nematodes and
fungi. Therefore soil fumigation with 1.3-dichloropropene or metham-sodium
is carried out in July once every 2 or 3 years. Both fumigants have a good
side effect on weeds. Therefore in practice it is often not necessary to
apply lenacil for weed control after a soil fumigation between two crops.
In case only nematodes are problematic, a fumigation with ethoprophos is
preferred by the growers, because no waiting period is needed before
planting. Since this chemical has no herbicidal effect lenacil (0.8-1.6
kg/ha) must be applied 2-3 weeks after planting (1).

3 Prolongation of the picking period in strawberry

Forcing strawberries by growing them under glass is an old method.
Nowadays the acreage of strawberry under glass comprises about 150 ha, of
which 70 ha are heated. Fumigation of the soil between cropping tomato (or
gherkin) and strawberry to kill the inoculum of soil borne diseases and
covering of the soil with opaque plastic film from the end of February
makes the application of herbicides mostly superfluous.

Another forcing method is covering strawberry plants grown in the open
under perforated plastic film in spring until they start flowering. The
acreage of this culture has increased during the last few years and
comprises, in 1985, about 200 ha. Covering starts already in the beginning
of March and then from time to time the plastic film is temporarily removed
to spray for control of diseases and weeds.

Retarding of the harvest period has already been possible for many
years by growing ever bearing varieties. In this culture the soil in the
plant rows is covered by a strip of plastic film to preserve soil moisture.
This makes weed control only necessary in the foot paths between the
strips. On the other hand practising strips is preferable, because the two
ever bearing varieties Ostara and Rapella are rather sensitive to
herbicidal sprays, showing yellowing of the leaf margins and some growth
retardation when sprayed (9). However, an increase of the acreage of the
ever bearing varieties (in 1985 about 150 ha) is hampered by the interest
of the growers for the late-summer culture of cold stored plants. Because
the growth period of such a culture takes only three months, lenacil can be
applied at the low rate of 0.8 kg/ha, provided no other strawberry culture
starts in the same year on the soil involved. In 1985 the acreage of late-
-summer culture of cold stored plants amounts to about 300 ha.

4 Strawberry varieties

In the Netherlands growers can make their choice from 19 varieties,
including 2 ever bearing ones, which are all officially recommended (4).
Besides the two ever bearing varieties Ostara and Rapella, the varieties
Bogota, Confitura and Korona are rather sensitive to a herbicidal spray.
Leaf margins on sprayed plants become yellow and the plants are more or
less retarded in their growth. For that reason it is advised to apply such
a herbicidal spray not before the first of October when the plants have
been in the production fields for two months. But then, hand-hoeing is
necessary during these two months.

5 Use of straw

For many years straw has been placed in the strawberry fields, mainly to keep the fruits clean, and besides it also retards the growth of the plants somewhat, resulting in a later harvest. Since this is of interest in case late-cropping varieties are grown, a later harvest may come at a time of higher prices. But, volunteer cereal plants germinate from seed in the straw and have to be controlled. That is done by "unlocking" the straw, giving it a treatment with anhydrous ammonia. To do so, the straw is collected in a kind of rick, covered with tarpaulin, after which the air is sucked out. Then the sucking is suddenly stopped and at the same time anhydrous ammonia is inserted. Since May 1985 also sethoxydim (0.46 kg/ha) has been registered for control of volunteer cereal plants and allowed to be sprayed until three weeks before harvest.

6 Hedgerow system in woody soft fruit

The change from bush trees to hedgerow growing in red currant and gooseberry is the most important one, leaving the bush tree growing system to black currant and blueberry only. The hedgerow growing system facilitates the application of paraquat (0.6-1.0 kg/ha) for control of weed vegetation. Of course, along a hedgerow it is easier to manipulate a protecting screen to prevent the shoots from being hit by paraquat.

Top fruit growers with an interest in growing soft fruit, are partial to mechanisation using four-wheeled tractor driven machinery. In that case a wider planting distance between the soft fruit hedgerows is necessary, as well as grass strips on the sticky clay soil, the soil type that prevails on holdings where woody soft fruit (excluding blueberry) is grown together with top fruit. Having grass strips, the growth of runner-producing perennials from these strips into the weed-free plant rows needs special attention. Spot treatment is sometimes necessary, because competition for moisture and food already occurs by the grass, and any additional competition is inadvisable.

7 Erect training of new canes in blackberry

Trailing canes in thorned blackberry varieties is a must, but in thornless varieties canes can be trained erect. Such a training results in fewer broken canes, which on average are more vigorous and less susceptible to frost damage. As a result the yield increases (5). Although under the bundle of trailed canes weeds are hardly able to germinate, deep-rooting weeds are sometimes problematic particularly in the first years of a plantation. In case of erect training of the canes these weeds are easier to hit by a herbicide and therefore easier to control.

In the last three years Amaranthus lividus has become a problem weed in some plantations in the south-west of The Netherlands (8). It has proved to be easily controllable by the use of diuron. But, to prevent a build-up of unwanted high residues in the soil, diuron should be applied at a rate of 1.6 kg/ha at the outside. As a consequence, it must not be applied in March together with simazine, but somewhere in May/June to achieve adequate control during the summer months. Because diuron is phytotoxic to the green parts of the blackberry plants, it has to be applied at that time with a protecting screen. Doing so, erect training of the canes makes it easier to apply the spray.

REFERENCES
1. ALOFS, W.J. (1983). Gewasbescherming vóór het planten van aardbeien. Groenten en Fruit 39, 42-43.
2. ALOFS, W.J. (1985). Verzorgen vermeerderingsveld aardbeien. Groenten en Fruit 40, 62-63.

3. ALOFS, W.J. (1985). Chemische of mechanische onkruidbestrijding in het voorjaar. De Fruitteelt 75, 300-301.
4. COMMISSIE voor de samenstelling van de Rassenlijst voor Fruitgewassen (1984). 17e Rassenlijst voor fruitgewassen 1985. Maastricht (Leiter-Nypels), pp. 200.
5. GEENSE, C. (1981). Ontwikkelingen in de bramenteelt. De Fruitteelt 71, 312-314.
6. GODDRIE, P.D. (1961). Chemische onkruidbestrijding in de fruitteelt. Wilhelminadorp. Mededeling Nr 3, Proefstation voor de Fruitteelt, pp. 12.
7. JOOSSE, M.L. (1984). Statistische gegevens van de Nederlandse fruitteelt. 2de herziene druk, Wilhelminadorp (Consulentschap in Algemene Dienst voor de Fruitteelt in de Volle Grond), pp. 103.
8. MOURIK, J. van (1985). Gewasbescherming. Deltafruit 2 (2), 16-17. Inserted in: De Fruitteelt 75 (10).
9. PEERBOOMS, H. (1974). Onkruidbestrijding bij doordragende aardbeien. De Fruitteelt 64, 595-596.

Chemical weed control in strawberry beds in Belgium

R.Lemaitre & P.Boxus
Station des Cultures Fruitières et Maraîchaires, Gembloux, Belgium

Summary

In Belgium the perennial strawberry plant is grown mainly as an annual.

- Planting for fruit production is carried out in early August using fresh
  transplants, or in June or July using cold-store transplants, and harvesting
  takes place in June or July of the following year.

  A number of special cultivation methods are also practised: early planting
  under glass or late planting outdoors, in which case the transplants
  are put outside in holding nurseries in early August, removed in winter
  during the dormancy period and kept in cold storage for a while. They are
  replanted under glass in December or January for early crops, or outside
  from May to July for late crops.

- The propagation beds receive fresh transplants in September or cold-stored
  transplants in March.
  In addition to a good knowledge of the flora to be killed, such diverse
  strawberry cultivation methods call for adaptation of weed control
  practices and a good understanding of how the herbicides work,
  their residual activity and their toxic effects on the subsequent crop.

1. Eradicating perennials

Before strawberry cultivation can begin, it is essential to destroy most of
the perennial weeds already present, e.g. Cirsium arvensis, Tussilago,
Equisetum, Agropyron repens, Rubus spp, Rumex spp, Lamium spp, Urtica urens.

If there are relatively few weeds, on cultivated land for example, they can be
eliminated locally using defoliant or hormone herbicides, e.g. Round-up,
2,4,5-T.

Planting on erstwhile meadows requires a greater amount of weed control, with
the best results being obtained as follows:

- October–November:ploughing of the meadow;
- from May to July: treating regrowth (15 cm) with a mixture of defoliant and
  hormone herbicides, e.g. Round-up + MCPA, paraquat or 2,4,5-T.

Mechanical preparation of the soil prior to planting kills plantlets which
have grown from seeds.

Black plastic mulching can be put down from May onwards when the ground is
ready: the plastic is perforated only at the time of planting in August, and
treatment of the inter-rows in the meantime is quite easy, e.g. paraquat +
simazine.

2. Outdoors

2.1. Annual cultivation

This is carried out on bare ground or using black plastic mulching.

2.1.1. Bare ground

The ground is kept clean by using root herbicides applied 8 to 10 days after planting in soil which is sufficiently moist and firmed down.

- Lenacil is long-lasting and offers the greatest guarantee for the strawberry plant.

Very effective against:	Poa annua,
	Matricaria recucita,
	Gnaphalium uliginosum,
	Stellaria media,
	Capsella Bursa-Pastoris.
Moderately effective:	Polygonum,
	Senecio vulgaris.
Not very effective:	Urtica urens,
	Lamium spp,
	Veronica spp,
	Viola arvensis,
	Cardamine hirsuta.

Doses: August, 12.5 - 15 g/are;
       April, 8 - 12 g/are.

When applied at the above dose in August lenacil may damage a subsequent crop such as gherkins, turnips or cabbages, especially if planted the following July.

In order to reduce the risks of residual activity a mixture of lenacil (10 g/are) + chloroxuron (15 - 17 g/are) should be applied in August, and followed by treatment with the same products in spring before mulching to destroy plantlets which have developed from adventitious seeds in the straw.

- Chloroxuron is too short-lived (two months).

Limited effectiveness against:	Poa annua,
	Gnaphalium,
	Lamium,
	Polygonum spp.

Doses: August, 35 - 40 g/are;
       April, 35 - 40 g/are.

In order to increase residual activity a mixture of lenacil (8 g/are) + chloroxuron (12.5 - 15 g/are) can be applied in April. Chloroxuron applied in spring to the young leaves of strawberry plants causes yellowing, but this soon disappears.

If the herbicide is applied late after planting - because of heavy rain or serious drought - adventitious seeds will have already sprouted, and the action of these products needs to be boosted by a defoliant - phenmedipham added to a wetting agent (8 - 11 g/are). Phenmedipham is effective up to

the 2-4 leaf stage against <u>Stellaria media</u>, <u>Chenopodium</u>, <u>Urtica urens</u>, <u>Galinsoga</u> and the cruciferae, but less so against <u>Viola arvensis</u> and <u>Veronica sp</u>; on the other hand, it is ineffective against the grasses <u>Polygonum sp</u> and <u>Matricaria</u>.
Certain strawberry plant varieties are sensitive to stronger doses of phenmedipham - Bogota, for example.

If the autumn treatment has not been successful, and after hand weeding of the rows, the inter-rows are treated at the end of winter with paraquat (8 - 10 g/are) using a shielded spray.

If there are a large number of grasses - <u>Echinochloa</u>, <u>Setaria viridis</u>, <u>Digitaria sanguinalis</u>, <u>Avena fatua</u>, <u>Lolium spp</u> and cereal regrowth (straw) - the following defoliants can be used:

> alloxydim sodium at 10 - 11 g/are)
> sethoxydim at 3.5 - 4 g/are   (+ 2% wetting agent
> fluazifop-butyl at 3 - 4 g/are.   (

These products should be applied before the strawberry plant blossoms, because after this period - in hot weather - growth is slowed down and the leaves turn yellow, especially with alloxydim.

2.1.2. Black plastic mulching

It is easy to keep the pathway soil clean by using simazine (7.5 - 10 g/are) or lenacil (15 - 17.5 g/are) before planting and prior to perforating the plastic. Paraquat (7 - 8 g/are) can be added if adventitious growth has already started.

At these doses there is usually no need to repeat the treatment in spring. If certain plants persist then phenmedipham, or even a further dose of paraquat, should be applied, but using a shielded spray this time.

In order to prevent gully erosion of the inter-rows in winter, especially on sloping ground, weak doses of simazine or lenacil, or even chloroxuron, are applied in August. The vegetation which develops at the end of October is destroyed in the spring by treating with paraquat + simazine: this weed mulch helps prevent pathway erosion and protects the soil from compaction.

2.2. Perennial cultivation

When the harvest is over, runners are destroyed by treating the inter-rows with diquat or paraquat (8 - 10 g/are), or dinoseb in an oily solution (3 - 5%).

At the end of September/early October fresh treatment is necessary to remove new runners. After manual weeding of the rows, mainly to eliminate runners, a general treatment of lenacil(7.5 - 10 g/are) - possibly together with phenmedipham (10 - 11 g/are) - should be applied. If <u>Convolvulus arvensis</u> or <u>Barbarea vulgaris</u> infestations have appeared in the meantime, a mixture of aminated MCPA (2.5 g/are) + aminated 2,4-D (3 g/are) can be applied in August. Chloropyralide (1.5 - 1.75 g/are) is effective against <u>Cirsium arvensis</u> and <u>Sonchus arvensis</u> when applied locally to well developed plants.

In spring lenacil (8 - 10 g/are) or chloroxuron (15 - 20 g/are) are applied. If tough annual dicotyledons such as <u>Amarantha</u>, <u>Viola arvensis</u>, <u>Veronica sp</u> or <u>Urtica urens</u> are present pendimethaline (12.5 - 15 g/are) can be added.

2.3. Holding nurseries

In this type of nursery the strawberry plants are planted close together in August (25 cm apart) with a view to transferring them under glass, or to cold storage for late crops, in December-January.

These beds are often treated, prior to planting, with DD (dichloropropane + dichloropropene) at a dose of 6 - 8 l/are or dazomet (4 - 6 kg/are), and then covered with plastic for a week. This treatment has a pronounced herbicidal effect, especially when using dazomet.

After planting, lenacil (8 g/are) or chloroxuron (30 g/are) is applied.

2.4. Propagation beds

Planting normally takes place in September in rows 1.6 m apart.

It is often enough to treat the rows with chloroxuron (35 - 40 g/are), even immediately after planting, by applying it to the base of the plants after watering. Adventitious weeds in the inter-rows are destroyed in spring by tilling or using a herbicide such as paraquat.

After the rows have been weeded, in May-June, nitrate fertilizer spread and the inter-rows given a final tilling, lenacil (12.5 - 15 g/are) is applied, or - if resistant weeds are feared - simazine (4 - 5 g/are).

Propagation beds planted in March are given the above treatment in April-May.

2.5. Late crops

Nursery transplants removed in January and then cold-stored are planted from May onwards. This cultivation method is very quick, taking just two months, but requires a lot of watering, which could allow the root herbicides to penetrate down to the roots of the strawberry plant. This is why lower doses are used: chloroxuron at 15 - 20 g/are and lenacil at 6 - 8 g/are.

3. Under glass

Use of herbicides is much reduced with this method.
The nursery transplants are planted in December, and opaque, white plastic mulching put down immediately after. Few weeds develop.

4. In tunnels

Cf. outdoors. Black plastic mulching usually used, which means that only the pathways require weeding.

5. Perforated plastic

Cf. outdoors.

Bibliography

Lemaitre, R., Le désherbage des cultures de fraisiers. Bulletin du fraisiériste 16 (5) 1970.

van Himme and Stryckers J., Le désherbage chimique du fraisier. Bulletin du fraisiériste 21 (1 - 2) 1975, Fruit Belge 53, No 410, 2nd quarter 1985.

Weeds

Latin name	French	English
AGROPYRON repens	Chiendent	Couch grass
AVENA fatua	Folle avoine	Wild oat
AVENA strigosa	Avoine strigeuse	Black oat
BARBAREA vulgaris	Herbe de Ste Barbe	Common wintercress
CAPSELLA Bursa-Pastoris	Bourse à Pasteur	Shepherd's purse
CARDAMINA hirsuta	Cresson de muraille	Hairy bittercress
CHENOPODIUM album	Chenopode blanc	Fat hen
CONVOLVULUS arvensis	Liseron des champs	Field bindweed
CRUCIFERAE	Crucifères	Crucifer
DIGITARIA sanguinalis	Digitaire sanguine	Large crabgrass
ECHINOCHLOA crus-galli	Pied de coq	Barnyard grass
GALINSOGA parviflora	Galinsoge	Smallflower galinsoga
GNAPHALIUM uliginosum	Immortelle sauvage	Marsh cudweed
LAMIUM spp	Lamier	Dead nettle
LOLIUM spp	Raygrass	Ryegrass
MATRICARIA recucita	Petite camomille	Wild chamomile
POA annua	Paturin annuel	Annual meadowgrass
POLYGONUM aviculare	Renouée des oiseaux	Knotgrass
POLYGONUM persicaria	Renouée persicaire	Lady's thumb
SETARA viridis	Sétaire	Foxtail or bristlegrass
SENECIO vulgaris	Seneçon commun	Common groundsel
STELLARIA media	Mouron blanc des oiseaux	Chickweed
URTICA urens	Petite ortie	Small nettle
VERONICA spp	Véronique	Speedwell
VIOLA arvensis	Pensée sauvage	Field violet or pansy
CIRSIUM arvensis	Chardon des champs	Canada thistle
SONCHUS arvensis	Laiteron des champs	Perennial sowthistle

Active ingredients			Commercial names

root-type

Simazine,	50 %	
Lenacil,	80 %	
Chloroxuron,	48 %	Venzar
Neburon,	60 %	Ténoran

defoliant

Phenmedipham,	15 %	Bétanal
Alloxydim sodium,	75 %	Fervin
Sethoxydim,		Fervinal
Fluazifop-Butyl,	25 %	Fusilade
Paraquat,	20 %	Grammoxone
Diquat,	19.5 %	Réglone
Dinoseb (DNBP),	25 %	
Glyphosate,	36 %	Round-up

hormone-type

aminated 2,4-D,	50 - 72 %	
aminated MCPA,	40 - 75 %	
Chloropyralide (3,6 dichloropicolinic acid) (3,6-DCP),	10 %	
Pendimethaline,	33 %	Stomp

Weeds and weed control in strawberries of Greece

C.N.Giannapolitis

'Benaki' Phytopathological Institute, Weed Department, Kiphissia, Greece

Summary

Strawberries in Greece are a minor crop (approximately 1000 ha) grown either under plastic cover in cheap greenhouse constructions, or uncovered in the field. Weeds become particularly troublesome if they are abundant during runner growth and flowering of the crop, which take place primarily in spring and secondarily in autumn. Some of the weeds frequently occurring at high densities in strawberries include *Echinochloa crus-galli, Setaria* spp., *Chenopodium album, Amaranthus* spp. and *Cyperus rotundus*. Weed control is based on mulching with black polyethelene plastic and/or small-grain straw and complemented with use of herbicides, hand weeding, hoeing and grazing of sheep. Two integrated weed management systems have evolved from combining these methods. The systems provide consistent weed control, especially in absence of perennial weeds and when coupled with a proper rotation of strawberries to other crops.

1. Introduction

Strawberries are a minor crop in Greece covering an acreage of not more than 1000 ha. They are grown in the southern as well as the northern part of the country in a number of well defined localities.

In about half the acreage strawberries are grown under plastic cover, in cheap greenhouse constructions, which allows earlier maturation of berries. Greenhouse-produced strawberries are brought to the fresh fruit market at a time (March – April) when other fruits are scarce, thus giving a good income to the growers. Field-produced (no plastic cover) berries are basically intended for processing; a proportion of them, however, is always forwarded to the fresh fruit market.

Strawberry beds are established from rooted hardwood cuttings late in summer or early in autumn. The soil is worked into ridges (60–80 cm) separated by furrows (30–40 cm). Strawberries are planted in two rows on the ridges.

Strawberry beds are kept for 2–3 years, if used for production of fresh-fruit-market berries. Otherwise, the beds may be kept for up to 6 years. Following this period, the beds are rotated to other crops or in some cases replanted with strawberries.

No systematic weed research has been conducted so far with strawberries in Greece. The discussion on weed problems and control techniques, which follows, is therefore based on observations and experience from field practice rather than on research data.

2. Weed problems

Weeds interfere with strawberry production in a variety of ways. Field observations in Greece clearly indicate that as a result of competition for moisture and nutrients weeds reduce yield, reduce quality (notably size) of berries and shorten productive life of the plantings. Furthermore, weeds hinder picking and enhance berry rotting.

In Greece detrimental effects of weeds become more pronounced if weeds are abundant during runner growth and flowering of the crop, which take place primarily in spring and secondarily in autumn. During the rest of the year, growth of the crop is very limited and practically not affected by the presence of weeds.

Weeds also become a serious problem soon after planting of a new strawberry bed. Unless control measures have been taken, weeds quickly overgrow young strawberry plants leading to failure in establishing the crop.

A great number of weed species can be found in strawberry fields of Greece. In a strawberry field, not well maintained near Katerini, for example, the weed flora consisted of 17 species (Table 1), including four perennials (*C. dactylon, C. rotundus, C. arvense* and *S. halepense*). Three of these species were present at high densities and four species at intermediate densities. The remaining species were present only as scattered plants.

Table 1 . Weed flora of a strawberry field in Katerini (N. Greece) in May

Weed species	Family	Relative abundance
Chenopodium album	Chenopodiaceae	high
Cynodon dactylon	Gramineae	high
Echinochloa crus-galli	Gramineae	high
Amaranthus retroflexus	Amarantaceae	medium
Arenaria sp.	Caryophyllaceae	medium
Cyperus rotundus	Cyperaceae	medium
Lolium rigidum	Gramineae	medium
Amaranthus lividus	Amarantaceae	low
Cirsium arvense	Compositae	low
Conyza bonariensis	Compositae	low
Lactuca seriola	Compositae	low
Lathyrus nissolia	Leguminosae	low
Raphanus raphanistrum	Cruciferae	low
Sorghum halepense	Gramineae	low
Verbena supina	Verbenaceae	low
Vulpia myurus	Gramineae	low
Xanthium spinosum	Compositae	low

Only certain species, however, are often found at high densities in strawberries. The 10 most important of these species are given in Table 2 arranged in a decreasing order of importance as defined by the frequency at which they prevail in strawberry fields. The *Setaria* species include *S. viridis* and *S. verticillata*. The *Amaranthus* species usually include *A. retroflexus, A. blitoides* and *A. albus* but occasionally other species as well (2). The *Polygonum* species include *P. aviculare, P. patulum* and possibly some other species.

Table 2. Weed species usually occurring at a high density in strawberry fields of Greece

Order of importance	Weed species
1	*Echinochloa crus-galli*
2	*Setaria* spp.
3	*Chenopodium album*
4	*Amaranthus* spp.
5	*Cyperus rotundus*
6	*Portulaca oleracea*
7	*Sinapis arvensis*
8	*Cynodon dactylon*
9	*Polygonum* spp.
10	*Convolvulus arvensis*

Some species which are more important in other crops may occasionally become serious weeds in strawberries. Examples of such weeds are *Avena sterilis, Phalaris* spp. (*P. minor, P. brachystachys* and *P. paradoxa*) and *Galium* spp. (mainly *G. spurium* but also *G. aparine* and *G. tricornutum*) (1,3).

3. Weed control measures

Soil fumigation with methyl bromide before planting is little used for strawberries in Greece. Mulching is widely used, instead. Besides weed control, mulching conserves moisture and keeps berries clean.

Black polyethylene plastic is used on the ridges and small-grain straw on the furrows in almost all strawberries grown for the fresh fruit market. Only straw is used in strawberries grown for processing.

Plastic mulching provides excellent long term weed control. Few weeds (mainly *Cyperus rotundus*) emerge by puncturing the plastic. Some other weeds may emerge from the holes around the strawberry plants if fitness is not close. Straw mulching provides only temporary suppression of weeds and usually needs combining with use of other measures to give satisfactory control. Volunteer cereals are a problem, whenever straw is used for mulching.

The grazing of sheep is quite often used for weed control in strawberries of northern Greece. Sheep enter the field just after the last picking of berries late in spring. By selectively grazing the existing weeds – they do not touch strawberry plants– sheep prevent weeds from producing seeds, thus reducing the chances of a weed build-up.

Table 3. Herbicides used on strawberries

Herbicide	Rate (Kg ai/ha)
Chlorthal-dimethyl	6.0–11.0
Paraquat	0.5– 1.0
Lenacil	1.1– 1.6
Alloxydim	1.1– 1.5
Diphenamid	5.0– 6.5
Napropamide	3.0– 4.5

Herbicides are used on strawberries in Greece mainly as a means to complement mulching. Herbicides used are given in Table 3 in order of decreasing usage. Chlorthal-dimethyl, diphenamid and napropamide applied before or after planting are useful in protecting young transplants from weed competition during establishment. The same herbicides and also lenacid applied on established beds some time from autumn to early spring provide residual pre-emergence control of weeds for the coming period of strawberry growth and flowering. Paraquat is applied as a spray directed to the furrows, just on the start of runner growth, to kill weeds before straw mulching is set. Alloxydium is used for selective grass weed control when weeds are at the 1-4 leaf stage.

4. Integrated weed management

Integrated weed management has found more practice in strawberries than in any other crop in Greece. This has undoubtedly been facilitated by the nature of the crop and the availability of a range of weed control measures one can choose from.

Two integrated weed management systems have evolved and are currently practiced by a number of growers.

The first system involves, on the ridges, plastic mulching prior to planting and hand weeding when needed thereafter. On the furrows, every year, a residual herbicide is applied in autumn and a contact herbicide early in spring followed by straw mulching.

The second system involves herbicide application in autumn or winter, hoeing followed by straw mulching in spring and sheep grazing in summer.

Depending on specific problems both systems may be complemented by additional measures (eg. selective sprays against grass weeds).

The first system as more expensive suits better the intensive cropping for fresh-fruit-market berries. The second system is preferred in the cropping situation where berries are intended for processing.

Results from both systems have been consistently good, providing that the field has been cleaned of perennial weeds prior to strawberry bed establishment. This is a necessity particularly if the second system is going to be followed. Furthermore, coupling the system with a proper crop rotation, and not replanting strawberries in the same field, also improves consistency in weed control.

REFERENCES

1. DAMANAKIS, M.E. (1982). A grass weed survey of the wheat fields in central Greece. Zizaniology 1 : 23-27
2. GIANNOPOLITIS, C.N. (1981). *Amaranthus* weed species in Greece : dormancy, germination and response to pre-emergence herbicides. Annls Inst. Phytopath. Benaki, 13 : 80-91
3. GIANNOPOLITIS, C.N. (1982). *Galium* weed species in wheat fields of Greece : spread, severity and effect on yield. Zizaniology 1: 5-10
4. WEED SCIENCE SOCIETY OF GREECE (1985). Guide to weed control. Recommendations for use of herbicides in strawberries (C.N. Giannopolitis). (In press).

Weed control in strawberries: Evaluation of new herbicides

D.Seipp

Versuchs- und Beratungsstation für Obst- und Gemüsebau-Vechta, FR Germany

Summary

In Germany Simazine, Lenacil and Chloroxuron are registered
for weed control in strawberries as pre-emergence residual
herbicides and Phenmedipham, as a post - emergence herbicide·
With regard to the Dutch varieties, which are widely grown
in Germany, there is a susceptibility to damage from Simazine and Lenacil (in particular cv. Bogota). Moreover there
are several weeds, which are not controlled by these herbicides, e.g. Veronica hederaefolia, Matricaria chamomilla,
Viola tricolor, Senecio vulgaris. They are becoming a problem in many strawberry fields. Therefore a number of herbicides (Napropamide, Diphenamid, Propyzamid, Pendimethalin
and Metamitron) were tested in a summer planting of the
cv. Bogota and were compared to Lenacil in their effect on
the strawberry plants as well as on weed control.

1. Introduction

The growing of strawberries in Germany has been a typical
horticultural crop with intensive cultivation on small areas
for a long time. Three factors have caused a thorough change
in this situation during the last 25 years :
1.) Breeding of new highly-productive varieties
2.) Development of effective fungicides against fruit-rots
3.) Introduction of herbicides for selective weed control in
   strawberries
These factors have been the reason for the expansion of the
production area from 3ooo ha in 196o up to about 6ooo ha in
1985 in Germany. At the same time there was a shifting of
strawberry production from many small growers to a large scale
production by farmers. The acreage of strawberries nowadays
covers often 1o and more hectares in a single farm. Only in the
Southwest of Germany (Baden) there is still an intensive production on small units where hoeing by hand still is integrated
into chemical weed control.
   Many of the large farmers are beginners in strawberry
culture, who expect a better profit from the strawberries than
from an agricultural crop. But very often, their enthusiasm is
suffocated by a solid layer of weeds which also kills the
strawberry crop and furthermore a lot of money and labour is

TABLE 1
EFFECTIVENESS OF SOME HERBICIDES AGAINST PROBLEM WEEDS IN STRAWBERRIES

Weed / Herbicide	Veronica hederaef.	Viola tricolor ssp. arvensis	Matricaria chamomilla	Senecio vulgaris	Galium aparine	Poa annua
Venzar (1,5 kg/ha) (8o % Lenacil)	−	−	±	±	−	+
Goltix (5 kg/ha) (7o % Metamitron)	+	±	+	±	±	+
Stomp (5 l/ha) (33o g/l Pendimethalin)	+	+	±	−	±	+
Kerb 5o W (1,5 kg/ha) (5o % Propyzamid)	+	−	−	−	−	+
Enide (14 kg/ha) (5o % Diphenamid)	+	+	+	+	n.t.	+
Devrinol (2,5 kg/ha) (5o % Napropamide)	+	−	+	+	n.t.	+

+ = effective   − = not effective

± = moderately effective   n.t. = not tested

100

wasted, which has been invested by this time.

The weeds do not only bring problems to the newcomer, but also to the grower who has cultivated strawberries for a number of years. These problems arise from the selection of certain weeds by the continual use of one specific herbicide. This leads very soon to an increased population of weeds difficult to control. In some cases the weeds have become resistent to herbicides as for instance Poa annua and Senecio vulgaris to Simazine. A further problem is the unsatisfactory tolerance of several new strawberry varieties to the registered standard herbicides. Most of the Dutch varieties show a low tolerance, in particular cv. Bogota, which reacts very sensitively to the herbicide Venzar (Lenacil). On the other hand this very late ripening variety has become of great importance in Northern Germany, because the growers want to extend the selling season into late summer in order to avoid the oversupply situation during mid-season, which normally means low prices.

That is why an experiment with different herbicides was started in 1983 at the Research and Advisory Station for Fruit and Vegetable Crops at Langförden (Oldenburg). A number of herbicides was chosen, which were known to show some tolerance in strawberries and which are sold in Germany. Furthermore, the herbicides should control a weed spectrum different from that of Venzar (Lenacil). None of them is registered for use in strawberries yet, but there is some interest by the chemical industry to get approval for them (See table 1).

2. Material and Methods

Runner plants of the strawberry variety Bogota were planted on 22.8.1983 at a distance of o,3o x 1,o m. Each plot covered an area of 2o m^2 with 4 replicates. The soil is a sandy loam (8% clay; 64% silt; 28% sand). The organic matter is approximately 4%. Some weeks after planting all plots were treated with 6 l/ha Betanal (Phenmedipham), and towards the end of the growing season the weeds which escaped this treatment were removed by hoeing. During August 1983 the young plants had to be irrigated twice, due to the lack of rain. The winter was relatively mild except for February, which brought some frost without snow. The spring was very dry, whereas from middle of May until the end of July the weather conditions were much cooler and wetter than in a normal year.

The chemicals tested can be classified as pre-emergence residual herbicides; two of them (Goltix and Kerb) having a post-emergence effect, too.
a.) Venzar (8o% Lenacil) with an application rate of 1,2 - 2,o kg/ha is still the standard residual herbicide in straw-berries in Germany with good control of many weeds with the exception of Viola tricolor, Senecio vulgaris, Galium aparine, Veronica hederaefolia, Solanum nigrum. It may be applied almost any time outside the flowering and harvesting period in young as well as established plantings. As mentioned above, there is some risk of phytotoxicity to some strawberry varieties.
b.) Goltix (7o% Metamitron) is registered for use in sugar-beet at   5 - 1o kg/ha. Only the lower rate of 5 kg may be used

TABLE 2
Effect of Different Herbicides on Weed Population in Strawberries (1.6.84)

Treatment	Degree of Weed-Cover %	Percentage of the Main Weed Species						
		Matricaria chamomilla	Capsella bursa-past.	Stellaria media	Veronica hederaef.	Senecio vulgaris	Poa annua	Miscellaneous Weeds
1. Untreated	20	50	20	10	10	5	5	0
2. Venzar (1,5 kg; 16.4.84)	12	30	30	25	10	5	0	0
3. Goltix (5,0 kg; 16.4.84)	10	10	30	25	15	10	0	10
4. Stomp (5,0 1; 30.11.83)	8	40	30	0	15	15	0	0
5. Kerb 50 W (1,5 kg; 30.11.83)	20	75	10	0	0	10	0	5
6. Enide (14 kg; 16.4.84)	5	50	30	0	0	20	0	0
7. Devrinol (2,5 kg; 30.11.83)	10	50	30	0	0	20	0	0

in strawberries on a well settled soil in spring or after
the first crop in summer. The variety Senga Sengana
generally shows good tolerance, but in a number of cases
damage to strawberry plants has been reported. In particular Goltix controls Matricaria chamomilla and Veronica
hederaefolia. In addition it is effective against a wide
spectrum of weeds not controlled by Venzar.
c.) Stomp (33o g/l Pendimethalin) is registered in Germany for
agricultural crops. Stomp may only be applied during dormant period in strawberries. There is a good control of
Veronica hederaefolia and Viola tricolor as well as many
other weeds. Matricaria chamomilla is only controlled to
some extent.
d.) Kerb (5o% Propyzamid) is registered in Germany for different fruit crops, lettuce and ornamentals. With a rate of
1,5 - 2,o kg/ha its main effect is on grasses and on Stellaria media, Polygonum persicaria and Urtica urens. Plants
of the composite family are not controlled. Kerb may only
be used during dormant period in strawberries, otherwise
plants will be extremely stunted.
e.) Enide (5o% Diphenamid) is not yet registered in Germany,
but in many countries it is used in strawberries. With
14 kg/ha there is a good control of Senecio vulgaris,
a problem weed with some herbicides. It may be applied at
similar times to Venzar.
f.) Devrinol (5o% Napropamide) has been recently registered in
Germany for weed control in rape. With 2,5 kg/ha there is
a good control of weeds in strawberries, especially Senecio
vulgaris, Veronica hederaefolia and Stellaria media but not
crucifers. The best time for application is before planting
as soil incorporated residual herbicide or during the dormant period on an established crop.
All these herbicides were compared to a plot weeded by hand.
This was done by hoeing three times, e.g. in early and late
autumn 1983 and in spring 1984, to avoid weed competition.

3. Results

3.1. Weed Control

The degree of weed cover was recorded on 1.6.1984 before
the strawberries were mulched with straw. Most weeds were found
in the control, which was hoed for the last time in early April
(see table 2). From the total degree of weed cover about 5o%
was due to Matricaria chamomilla, 2o% to Capsella bursa-
pastoris and 3o% to other species. A similar result was found
after the application of Kerb, the main weed being Matricaria
chamomilla. Very few weeds were found in the plots with Stomp
and Enide followed by Devrinol and Goltix. Enide and Devrinol
controlled Stellaria media, Veronica hederaefolia and Poa annua
well, whereas Goltix and Stomp showed no specific effectiveness.

3.2. Phytotoxicity

The application of Kerb, Stomp and Devrinol was made on

TABLE 3
INFLUENCE OF DIFFERENT HERBICIDES ON THE YIELD
OF STRAWBERRY, CV. 'BOGOTA'

Treatment	Average Yield per plot (20 m^2) kg	% of control
1. Untreated	25,3	1oo
2. Venzar (1,5 kg/ha; 16.4.84)	27,8	113
3. Goltix (5,o kg/ha; 16.4.84)	34,9	138
4. Stomp (5,o l/ha; 3o.11.83)	33,1	13o
5. Kerb 5o W (2,o kg/ha; 3o.11.83)	24,9	98
6. Enide (14 kg/ha; 16.4.84)	32,3	127
7. Devrinol (2,5 kg/ha; 3o.11.83)	2o,1	79

3o.11.1983 when the plants were dormant. The other herbicides were sprayed on 16.4.1984. Symptoms of phytotoxicity were found only in a few cases in the springtime. Enide showed some yellowing of the leaves, which disappeared soon. There was also some very faint yellowing of the edges of the leaves after the application of Goltix. All other treatments had no visible damage.

3.3 Yield

Besides visible phytotoxicity the yield might be a measure of crop tolerance to the herbicide. Bogota is a very heavy cropping variety. As a result of the somewhat late planting date the yield of the untreated check is not very high with 25,3 kg per plot (about 43o g per plant). This may be due to temporary weed competition or to cultivation, which might have damaged some roots. Venzar did not affect the yield very much, there was slight increase in yield of about 1o% compared to the untreated plot. The highest yields, however, were found with Goltix and not much less with Stomp and Enide. All these plots showed an increase of about 3o% (see table 3). The lowest yield was recorded for Devrinol, where the plants yielded 2o% less than in the untreated control.

4. Discussion

From the results reported here, it seems to be possible to use most herbicides mentioned above in strawberries. With some of them there is no greater risk than in applying Venzar, which may lead to some damage onthe leaves. It is not evident why Venzar showed no phytotoxicity in spring 1984 nor in spring 1985 when the trial was repeated. Perhaps the application on a well settled soil is not as critical as on a soil which was recently cultivated. In addition to that the spring applications were not preceded by an application in the previous summer. But nevertheless there might have been some depression in growth for the yield was not as high as with some other herbicides. The actual weed control was satisfactory but with the growers there are a lot more problem-weeds than were evident in the trial plots.

Many ofthese problem weeds can be controlled by both Stomp and Goltix. Stomp will not cause phytotoxicity when it is applied during dormant period. Compared to this the use of Goltix may lead to a greater risk because the chemical is easily translocated into the rhizosphere by rain or irrigation. In this case several losses have been reported, even in established plantings. On the other hand Goltix shows good weed control and the highest yields in these trials.

Kerb has shown a good control of Stellaria media, Veronica hederaefolia and Poa annua, but the degree of weed cover of the plots was as high as in the check due to the strong growth of Matricaria chamomilla, which is not controlled. Even if there was no visible damage to the plants, the yield seemed to be affected. That is why it is more advisable to use Enide, which gives agood overall control of the weeds with a spectrum similar to Kerb. In addition to that there is no reduction in yield as it is with Kerb.

Though there was no visible damage to the plants, Devrinol led to the lowest yield. According to these results, it cannot be recommended for strawberries.

Bogota was chosen for this trial as a highly susceptible variety in tolerance to herbicides. Therefore the results may be somewhat unexpected for there was no obvious damage neither by Venzar nor by other herbicides. To ensure these findings in respect to phytotoxicity it is necessary to repeat the trial in order to meet with other weather conditions which might be of importance.

Even if Bogota is known to be very susceptible to Venzar and Simazine, this does not mean that it is necessarily susceptible to other herbicides. Therefore, these results may not be applied to other varieties, which might be susceptible to these new herbicides even if they are tolerant to Venzar. If another trial with a collection of standard varieties gives such a promising result similar to that obtained in this trial, the chemical industry could be asked to apply for the registration of these herbicides for use in strawberries. This would be of great interest to growers, who would then be able to choose between several herbicides in order to avoid a further selection of weeds, which are not controlled by Venzar.

REFERENCES

1. KOLBE,W.(1984). Zehn Jahre Versuche mit Goltix zur Unkraut-bekämpfung im Zucker- und Futterrübenanbau. Pflanzenschutz-Nachrichten Bayer 37,424-5o5.
2. LEMBRICH,H.(1978). Goltix, ein Herbizid mit hoher Selekti-vität für den Zucker- und Futterrübenanbau. Pflanzenschutz-Nachrichten Bayer 31,197-228.
3. SOOSTEN,R.VON(1983). Ergebnisse eines Unkrautbekämpfungs-versuches in Erdbeeren. Obstbau 8,233

Weed control in strawberries: The use of soil acting residual herbicides

N.Rath

Soft Fruit Research Station, Clonroche, Ireland

Summary

Several residual herbicides were compared for the control of germinating weeds in newly planted and established strawberries in trials carried out at the Soft Fruit Research Station, Clonroche. The herbicides were compared with the standard lenacil and include pendimethalin, napropamide, metamitron, ethofumesate, oxadiazon and oxyfluorfen. Winter and early spring applications of pendimethalin and of napropamide gave excellent control of annual weeds without causing plant damage. Napropamide also maintained plots free of Ranunculus repens and Trifolium repens. Mixtures of pendimethalin and napropamide also gave excellent weed control. Metamitron and ethofumesate did not cause plant damage but control of annual weeds was not satisfactory. Although winter applications of oxadiazon and oxyfluorfen gave good weed control, plant damage was severe.

Introduction

The standard residual herbicides for use on newly planted and established strawberries, viz lenacil and simazine suffer from a number of defects. The residual life of lenacil is short while simazine can not be used on newly planted strawberries unless the roots have been dipped in activated charcoal before planting. These herbicides do not control the full range of common weeds. A number of newer herbicides have shown promise on strawberries. Clay (1980) found that strawberries tolerated root applications of pendimethalin, ethofumesate, oxadiazon and oxyfluorfen but foliar applications of oxadiazon and oxyfluorfen caused severe damage. He also found that both root and foliar applications of metamitron caused slight damage. Davidson and Bailey (1980) found that spring applications of pendimethalin was safe under British field conditions. Metamitron had been used safely at Clonroche in previous trials (Rath and O'Callaghan, 1982). Lawson and Wiseman (1980) found that ethofumesate was promising for the control of annual weeds in Cambridge Favourite in Scotland. Clay (1980 b) found that oxadiazon could be applied safely to established strawberries during December. Napropamide is recommended for use on strawberries in the United States of America (Weed Control Manual 1984) and has also given promising results in Britain (Mathews and Wright, 1984).

Materials, Methods and Results

All the trials were carried out at the Soft Fruit Research Station at Clonroche. The soil is a clay loam and is typical of the soil on which most Irish strawberries are grown. It contained in the 0 - 0.15 m depth approximately 18% coarse sand, 11% find sand, 41% silt and 30%

clay. The standard cultivars Cambridge Favourite and Cambridge Vigour were used in these trials. The strawberries were planted on ridges which were 0.25 m high and 0.87 m apart. Cambridge Favourite plants were spaced 0.4 m apart while Cambridge Vigour was spaced 0.45 m apart. Cambridge Vigour was grown as spaced plants while Cambridge Favourite was allowed to form matted rows. Ridges were normally made up at least a month before planting and were sprayed off with paraquat immediately before planting. Experimental herbicide treatments were applied in winter or early spring. If necessary, trial areas were cleaned up by hand each August and were treated with simazine overall. Runners were controlled by interrow spraying with either paraquat or dinoseb-in-oil.

Experimental treatments were applied at approximately 1000 l/ha using an Azopropane sprayer fitted with Birchmeier Helico Sapphire nozzles.

Experiments 1 and 2

Strawberries, cv. Cambridge Favourite (experiment 1) and Cambridge Vigour (experiment 2) were planted on November 12, 1981. Most of the treatments listed in Table 1 were applied on December 3, 1981, while treatments involving the use of oxadiazon were applied on February 5, 1982. The treatments were again applied to the same plots on March 14, 1983, on February 15, 1984 and on February 21, 1985. On these occasions treatments with oxadiazon alone and with oxadiazon plus lenacil were replaced by pendimethalin plus propachlor and with pendimethalin plus napropamide respectively. The whole trial areas were cleaned up by hand and treated with simazine each autumn.

During March 1982 oxadiazon treatments caused severe scorching of the foliage of both strawberry cultivars. During April and May the damaged plants recovered and at the end of the growing season there was no difference in plant size on the different treatments. After the initial scorching caused by oxadiazon treatments no subsequent herbicide damage was observed. During May 1983 some slight transient stunting was observed on Cambridge Vigour plots treated with pendimethalin. Plant growth during 1982, 1983 and 1984 was excellent. No treatment affected crop yield in 1983 or 1984 (Table 1).

During August 1984 some patches of Cambridge Favourite were infested with Tarsonemid mite and the trial areas were treated with endosulfan. During spring 1985 severe symptoms of June Yellows showed up in the Cambridge Favourite with up to 30% of the plants showing some yellowing. The severity of the symptom did not seem to be affected by herbicide treatment.

Only small numbers of annual weeds germinated in the trial area and differences in the level of weed control on the different plots were slight. During 1982 and 1983 best weed control was obtained with pendimethalin and with napropamide treatments. During summer 1984 the plots treated with pendimethalin alone were lightly infested with Senecio vulgaris. Although the plots were cleaned up by hand each autumn, by spring 1985 the Cambridge Favourite trial area (experiment 1) was unevenly infested with perennial weeds. Trifolium repens occurred in patches throughout the trial area but did not occur on any plots treated with napropamide. During spring 1985 some S. vulgaris occurred on all plots treated with pendimethalin alone.

TABLE 1. Residual herbicides on strawberries 1981 - 1985

| | | Crop yield (tonnes/ha) | | | |
| | | C. Favourite | | C. Vigour | |
Treatment	Dose kg/ha	1983	1984	1983	1984
Lenacil	1.8	28.9	31.7	23.5	22.5
Pendimethalin	2.0	27.5	30.2	23.4	22.7
Pendimethalin	1.0				
+ lenacil	+0.9	27.3	31.4	24.9	22.7
Oxadiazon*	2.0	26.4	31.8	24.2	23.1
Oxadiazon*	2.0				
+ lenacil**	+1.8	27.1	31.8	23.1	22.0
Napropamide	4.5	27.2	32.0	25.1	23.2
Napropamide	2.3				
+ lenacil	+0.9	29.2	32.0	25.8	23.5
Metamitron	5.3	28.3	32.5	-	-
F test		NS	NS	NS	NS

* Pendimethalin at 1.0 kg/ha + propachlor at 2.9 kg/ha substituted from March 14, 1983 onwards.

**Pendimethalin at 1.0 kg/ha + napropamide at 2.3 kg/ha substituted from March 14, 1983 onwards.

Experiment 3
        Strawberries, cv. Cambridge Favourite were planted on April 13, 1983. The treatments listed in Table 2 were applied on May 17, 1983 and were reapplied to the same plots on February 13, 1984 and on February 22, 1985.
        Oxyfluorfen caused severe scorching of the strawberry foliage during late May, 1983. During June and July the plants on those plots recovered and by autumn these plants were only slightly smaller than those on the other plots. Slight leaf scorch also followed the applications of oxyfluorfen in 1984 and in 1985. No other herbicide treatment caused symptoms of herbicide damage in 1983, 1984 or 1985.
        During 1984 part of the trial area showed symptoms of red core and as a result plant growth and crop yield were uneven. While crop yield in 1984 was not significantly affected by herbicide treatment, the lowest yield was obtained on the plots treated with oxyfluorfen.
        During spring 1985 the strawberry plants   in two blocks of this trial showed severe symptoms of June Yellows. The remaining two blocks which were planted using a different source of propagation material showed only slight symptoms. The severity of the June Yellows did not appear to be affected by herbicide treatment.
        Germination of annual weeds in the trial area during 1983 and 1984 was sparse. Galium aparine was the dominant weed in both years. This weed was best controlled with oxyfluorfen. Pendimethalin at 2.0 kg/ha and napropamide also gave good control of G. aparine. Patches of perennial weeds gradually became established in the trial area during 1983 and 1984. By spring 1985 the trial area was severely infested with the perennial weeds Ranunculus repens, Potentilla reptans, Trifolium repens, Cirsium arvense and Carex spp. Best control of perennial weeds was obtained on plots treated with napropamide alone

or with napropamide plus pendimethalin. Only P. reptans occurred on those plots. Napropamide appeared to prevent establishment of R. repens, T. repens and C. arvense.

TABLE 2.   Residual herbicide on C. Favourite 1983 - 1985

Treatment	Dose kg/ha	Crop yield (tonnes/ha) 1984
Lenacil	1.8	18.5
Pendimethalin	2.0	22.5
Napropamide	4.5	19.8
Metamitron	3.5	19.7
Ethofumesate	1.0	19.4
Napropamide	2.3	
+ pendimethalin*	+1.0	22.5
Oxyfluorfen	0.5	17.2
F test		NS

*Chlorthal dimethyl at 4.5 kg/ha plus propachlor at 4.8 kg/ha applied in experiment 3 on May 18, 1983.

Experiment 4
    Strawberries, cv. Cambridge Favourite were planted on October 25, 1983. The treatments listed in Table 3 were also applied to experiment 4 on February 13, 1984 and were reapplied to the same plots on February 22, 1985. Oxyfluorfen treatments severely scorched the foliage of the strawberries during March 1984. Subsequently the plants recovered well and at the end of the growing season there was little difference in plant size on the different plots. Oxyfluorfen treatments again caused severe scorching of the strawberry foliage and stunting the plants during spring 1985.
    Few annual weeds occurred in the trial area during 1984. The most prevalent weed was G. aparine. This weed was best controlled with napropamide and pendimethalin. During spring 1985 occasional patches of perennial weeds established in the trial area. R. repens was the most widely occurring weed. This weed failed completely to establish on any plot treated with napropamide.

Experiments 5 and 6
    Strawberries, cv. Cambridge Favourite (experiment 5) and cv. Cambridge Vigour were planted in adjacent areas at Clonroche on November 5, 1984. The treatments listed in Table 2 were also applied to these trials on February 27, 1985. During March 1985 severe scorching of strawberry foliage occurred on the plots treated with oxyfluorfen. The strawberry plants on those plots remained stunted during May 1985. Plant growth in both trials was uneven in spring 1985 with patches of stunted plants on many plots. The stunting appeared to be more prevalent on Cambridge Vigour plots treated with pendimethalin.
    During spring 1985 Fumaria officinalis was the dominant weed in both

trial areas. Ethofumesate failed completely to control this weed. Napropamide, metamitron and lenacil also failed to control F. officinalis. Best control of F. officinalis was obtained with pendimethalin and with oxyfluorfen. Volunteer winter wheat plants were established throughout the trial area at the time of spraying and were not controlled by any of the spray treatments. Galeopsis tetrabit established and grew well on the plots treated with napropamide alone. This weed was controlled by all the other herbicides. During May 1985 some R. repens established on all plots except those treated with napropamide.

Discussion

Experiments 1, 2, 3 and 4 were established in land which one to two years previously had grown either blackcurrants or raspberries for several years. Control of annual weeds had been excellent in those crops and as a result germination of annual weeds in the trial areas was sparse. Annual weeds did not compete strongly with the strawberries at any stage and as a result crop yield in experiments 1, 2 and 3 in 1983 and 1984 was not affected by the level of weed control.

Because of the high organic matter content of Irish soils and also because of the relatively high clay content of our principal strawberry growing soils, lenacil is regarded as a very safe herbicide in Ireland. However, it has a short residual life, does not control the full range of commonly occurring annual weeds and is ineffective on dry soil conditions. Lenacil is, therefore ineffective in many situations. Although in the present series of trials lenacil controlled annual weeds well, it allowed seedlings of R. repens to become established and this weed is now a problem in three of the longer established trials.

Pendimethalin is a residual herbicide recommended for use on established and newly planted strawberries between late autumn and early spring. When used during this period in the present series of trials pendimethalin treatments did not at any stage adversely affect crop yield. Pendimethalin did cause slight stunting of Cambridge Vigour on one occasion in experiment 2. Damage which may have occurred in experiments 5 and 6 was obscured by uneven plant growth. Pendimethalin has been used widely on autumn planted strawberries during the last two years and only slight, occasional stunting has been recorded (Burke, J. 1985, personal communication).

Pendimethalin gave excellent control of most common annual weeds occurring in the trial areas. However, it did not control S. vulgaris, a very important weed known to be resistant to pendimethalin. Clumps of R. repens and T. repens became established in plots treated with pendimethalin alone in experiments 1, 3 and 4 and this is probably a reflection of its inability to control seedlings of these species seen in experiments 5 and 6.

Napropamide is also a residual herbicide cleared for use on established strawberries. Because it is broken down by sunlight it is recommended for use between November and February only. While good results were obtained from a May application of napropamide in experiment 3, this was associated heavy rainfall immediately before and after application of the herbicide.

In the present series of trials napropamide was used on newly planted and established Cambridge Favourite and Cambridge Vigour over a four year period. Napropamide did not on any occasion adversely

111

strawberry growth or crop yield. Mathews and Wright (1984) recorded some reduction in vigour of newly planted Hapil and also slight reductions of yield of newly planted Hapil and newly planted Cambridge Favourite. These reductions are not likely to be important in an Irish situation where strawberries are usually not cropped in the year of planting.

The weed control obtained with napropamide in the present series of trials was excellent. Napropamide was particularly useful in preventing the establishment of R. repens and T. repens. In addition napropamide gives excellent control of both ordinary and simazine resistant S. vulgaris, a weed not controlled by pendimethalin. In experiments 5 and 6 napropamide failed to control F. officinalis and G. tetrahit, two weeds listed by the manufacturer as being susceptible to napropamide.

Mixtures of half strength napropamide and pendimethalin gave consistently good weed control without plant damage in the present trials. Because of their complementary weed control spectra and long residual life, mixtures or sequences of half strength napropamide and pendimethalin are probably the most effective herbicide treatments for use on autumn planted strawberries under Irish conditions. However, because of its expense it is likely that napropamide could only be justified as a band treatment along the plant row while a cheaper herbicide like simazine could be used in the alleyway. It is difficult to justify overall application of napropamide to established strawberries unless a severe problem with S. vulgaris occurs.

Both oxadiazon and oxyfluorfen treatments caused severe scorching and stunting on all occasions when applied to strawberries in these trials. Although the strawberries sometimes made a complete recovery, these treatments are too severe and could not be recommended for use on strawberries.

Metamitron was tested in five of the six trials and on no occasion did it cause damage to the strawberry plants. Previous trials at Clonroche also showed that metamitron was safe on strawberries (Rath and O'Callaghan, 1982). Metamitron has a shorter residual life than napropamide or pendimethalin. Its use on strawberries is likely to be confined to spring planted strawberries either as a replacement for lenacil or as a component of a low dose mixture with phenmedipham.

Although ethofumesate did not cause damage to strawberries in the present series of trials, in previous trials damage was severe, particularly to Cambridge Vigour (Rath and O'Callaghan, 1982). The usefulness of ethofumesate is confined to the control of established clover in strawberries.

References

Clay, D.V. (1980 a).   The effect of application timing and formulation on the tolerance of strawberries to oxadiazon. Proceeding 1980 British Crop Protection Conference - Weeds, 337 - 344.

Clay, D.V. (1980 b).   The use of separate root and shoot tests in the screening of herbicides for strawberries, Weed Research 20, 97 - 102.

Davidson, J.G. and Bailey, J.A. (1980). The response of strawberries
to spring applications of pendimethalin.
Proceedings 1980 British Crop Protection
Conference - Weeds 329 - 386.

Lawson, H.M. and Wiseman, J.S. (1980). Herbicide programmes for spring -
planted strawberries. Proceedings 1980 British
Crop Protection Conference - Weeds 353 - 360.

Mathew, P.R., and Wright, H.C. (1984). Weed control in strawberries
with winter applications of napropamide. Aspects
of Applied Biology 8, 1984 Weed control in fruit
crops 123 - 131.

Rath, N. and O'Callaghan, T.F. (1982). A comparison of residual
herbicides on newly planted and established
strawberries in Ireland. Proceedings 1982 British
Crop Protection Conference - Weeds 267 - 273.

Weed Control Manual 1984 and Herbicide Guide. Published by Ag Consultant
and Fieldman pp 340.

Experiments on weed control in strawberries and nursery plants

G.Marocchi

Osservatorio Regionale per le Malattie delle Piante, Bologna, Italy

Summary

The strawberry nursery presents many problems of weed control, as it is planted in a fertile soil and must be cultivated for 12/13 months long. In such a long time weeds can grow very rapidly and require many expensive mechanical treatments. The tests carried out during many years gave valid and useful results. The best results were obtained applying trifluralin, dinitramine, trifluralin + lenacil, dinitramine + lenacil, oxadiazon in pre- and in post-transplanting. The above mentioned chemicals are not sufficient to grant a complete control during the whole cultivation time. So, even one or more mechanical treatments are necessary to provide excellent weed control during the cultivation period.

We can report similar observations and results about the weed control in many other plants in the nursery and in soft fruit generally.

1.1 Results of weed control in fruit nurseries and ornamental plants

Weed control in nurseries of fruit-bearing plants is an extremely interesting agricultural technique and it is necessary to have an acceptable solution against the high costs involved in manual or mechanical weed control. It is as much as ever complicated by a multiplicity of factors: the different plant species, the various methods used on the same nursery - plant production by seed, in vitro and other techniques.

To reach a solution or to throw some light on the problem a programme of trials was started some years ago, which have already given a number of conclusive results. It is logical that it is necessary to proceed with the greatest caution, on account of the high value of the material which has to be treated.

In 1982, three weed control trials were conducted in nurseries already planted. These were not aiming for a definitive solution but were only observational to test the possibilities and the efficacy of some treatments with herbicides.

In 1983-85, using the information from previous trials, different applications were made to large areas and in all cases previous results were confirmed. Particularly could be seen the perfect repeatability of trifluralin, which applied pre-transplanting, was always highly selective. Also lenacil used in pre- and in post-transplanting periods has given the same results.

A limited selection of suitable products was drawn up because the results of other trials showed that there were many phytotoxic herbicides.

The treatment was made on April 5 after the plants had started vegetative growth. Treatment at that stage did not demonstrate any sign of phytotoxicity with any products used. This trial confirmed yet again the optimal selectivity of lenacil (Venzar) and also its possibilities for weed control. Venzar has many limitations in weed control if used alone, particularly against Amaranthus and Polygonum aviculare. Herbicidal activity improves distinctly with the addition of products like trifluralin, ethalfluralin etc. (1- 3.5 kg per ha) with nearly total control of weeds for a long period of time. The product linuron and pendimethalin (Panter or Inex) has given also an optimal result and therefore this product is to be considered further for this sector of weed control. Anyhow it is necessary that the fruit plants are not in the vegetative stage and do not have many young parts (newly expanded leaves or soft branches). This was demonstrated in other trials with Panter and the phytotoxic result was shown when applied on green vegetation.

Venzar used alone or in mixture with products like trifluralin, ethalfluralin, oryzalin and others has never shown any sign of phytotoxicity even if applied to very small plants (or larger plants with treatments applied specifically to folaige in a completely vegetative stage). This is a fact of extreme interest which leads us to believe that the problem of weed control in fruit plant nurseries is solved or at least reduced to manageable proportions. Also to be taken into account is the fact that during the trials aimed at a solution to this problem a number of different mini-trials were conducted on various plants (peaches, plums and others) with Venzar alone or in the above-mentioned mixtures. In these trials also no phytotoxicity occurred. Extending these trials also to the forestry and ornamental plants sectors we got the same positive results.

2.1 First results of trials of herbicides in ornamental cultures and fruit trees - 1983-1985

The method described above of chemical weed control in fruit tree nurseries was made in 1983 and was used not only in small trials but on large areas. Every time with the confirmation of the acquired elements, this means with optimal results.

Applications were made under pears, apple, peaches, plums, etc. With peaches a trial was made in the nursery with plants nearly 1m high and with large leaves. Trifluralin, ethalfluralin and lenacil applied with the tested technique gave the expected results without the minimum sign of plant injury or phytotoxicity.

2.2 Trials with herbicides on Laurus

Trifluralin, ethalfluralin and mixtures of these herbicides with lenacil etc. were applied on plants of Laurus (40-60cm high). In this sector also the results obtained with fruit trees were confirmed. This suggests that it is possible with these herbicides to extend their use to a large number of ornamental plants, forest-tree nurseries and to soft fruit generally. These were the objectives in the following trials.

3.1 Weed control in the strawberry nursery

A strawberry nursery generally presents some important problems concerning weed infestations: the soil is in fact very fertile, thus favouring the rapid establishment and development of weeds; furthermore the small plants are planted in the nursery for 12-13 months long (from September - when they are planted until September/October of the following year). During such a long period, infestation with weeds can easily develop.

Weed control in a strawberry nursery requires many mechanical treatments; these are not only expensive but often damage the crop plants. Under these circumstances studies on herbicides and application methods are an urgent necessity. The chemical products must be chosen from among those that are most effective for weed control and yet not damaging to small strawberry plants; the strawberry plants spread through their stolons and the roots, which give rise to the new small plants, have to find a soil in which weeds are chemically controlled by very selective products.

To solve this difficult problem we have carried out for many years a close set of tests which are now giving the expected results. It was evident that the problem required the integrated use of both chemical and mechanical methods. The chemical and mechanical treatments, properly completed and followed one upon the other, grant a perfect weed control throughout the life of the plantation without damaging the strawberry plants. To obtain this result the tests carried out during 2 years, included in fact chemicals and mechanical or manual weeding. The chemical treatments were carried out in autumn at the time of planting the strawberries and at the time of transplanting the small plants. A randomized block design was used with four repetitions for each application.

Table 1 shows the products and the rates applied. It also indicates the percentage level of weed control achieved (average of four replicates) compared with the untreated controls. The assessments were carried out on April 20, 1982 and in September 1982.

3.2 Discussion

The trial described above which follows other trials carried out in the pst years, has given very interesting rersults, which may be of help for applicztyions in the field. To control weeds in strawberry nursery beds (runner plantations), applications have to begin at the time of planting and both pre- and post-planting treatments are necessary; these applications must be followed by other chemical applications.

3.3 Conclusion

Trifluralin, applied as a pre-planting treatment at the rate of 0.53 kg/ha gives good control of dicotyledonous, weeds, although it does not control well Alopecurus myosuroides. Higher rates of trifluralin as a pre-plant treatment have partially controlled Alopecurus myosuroides.

After the fall application with trifluralin and after a mechanical treatment of the soil, effective applications of mixtures of trifluralin + lenacil or dimitramine + lenacil were carried out in April. After these applications the nursery was scarcely infested until the end of the cultivation.

Applications of benfluralin or trifluralin + linuron have given similar results to trifluralin applied under the same conditions. These applications were followed by further treatments with trifluralin + lenacil giving complete weed control until the end of the cultivation.

Table I - Treatments and results

FIRST PART OF THE TRIAL: from the establishment of the nursery (in October 82 to 20thApril 83

	Treatments	Dose a.i Kg/ha	Control % 20.4.
1	Trifluralin	0.53	50
2	Trifluralin	0.53	50
3	Trifluralin	0.53	50
4	Trifluralin	0.53	50
5	Trifluralin	0.53	50
6	Trifluralin	0.80	65
7	Trifluralin	0.80	65
8	Benfluralin	1.20	71
9	Trifluralin + Linuron	0.70+0.25	76
10	Napropamide	2.50	96
11	Napropamide	2.50	96
12	Diphenamid	4.=	92
13	Diphenamid	4.=	92
14	Dinitramine	0.50	68
15	Dinitramine	0.50	68
16	Oxadiazon	1.50	98
17	Oxadiazon	1.50	98
18	Oxadiazon	1.50	98
19	Trifluralin	0.88	70
20	Dinitramine	0.75	75
21	Lenacil + Trifluralin	1.20+0.66	97
22	Lenacil + Trifluralin	1.20+0.66	97
23	Lenacil + Dinitramine	1.20+0.50	97
24	Lenacil + Dinitramine	1.20+0.50	97
25	Mechanical weeding	- - -	100
26	Untreated	- - -	0

SECOND PART OF THE TRIAL: from mechanical weeding (20thApril 83 to September 83

	Treatments	Dose a.i Kg/ha	Control % 15.9
1	Chlorthal + Cloroxuron	7.50+2.50	
2	Chlorthal + Clorpropham	7.50+1.25	
3	Lenacil + Trifluralin	1.20+0.66	1
4	Lenacil + Dinitramine	1.20+0.75	1
5	(No other treatment)	- - -	
6	Trifluralin	0.66	1
7	Dinitramine	0.75	
8	Lenacil + Trifluralin	0.80+0.44	
9	Lenacil + Trifluralin	0.80+0.44	1
10	Lenacil + Trifluralin	0.80+0.44	
11	(No other treatment)	- - -	
12	Lenacil + Trifluralin	0.80+0.44	1
13	(No other treatment)	- - -	
14	Lenacil + Dinitramine	0.80+0.50	1
15	(No other treatment)	- - -	
16	Lenacil + Trifluralin	1.60+0.80	1
17	Lenacil + Dinitramine	1.60+1.=	1
18	(No other treatment)	- - -	
19	Lenacil + Trifluralin	1.60+0.88	
20	Lenacil + Dinitramine	1.60+1.=	
21	Lenacil + Trifluralin	0.80+0.44	1
22	Lenacil + Dinitramine	0.80+0.50	1
23	Lenacil + Trifluralin	0.80+0.44	1
24	Lenacil + Dinitramine	0.80+0.50	1
25	Mechanical weeding	- - -	10
26	Untreated	- - -	

Napropamide or diphanamid applied in autumn gave good weed control but were phytotoxic to strawberries.   Napropamide did not control _Veronica_ spp and diphenamid did not control _Fumaria officinalis_.

The   results   obtained with the autumn application of dinitramin   were similar to those obtained with trifluralin: _Alopecurus myosuroides_ was not well controlled although trifluralin was effective on dicotyledons.   An application with lenacil + dinitramin was carried out in spring: it gave excellent   control   of   the   weeds   present   without   phytotoxicity   to strawberries.

Effective control was obtained with oxadiazon, applied in Fall.   This product may be applied as a post-planting treatment giving excellent control of mono and dicotyledons.   An oxadiazon application in fall or in winter may be sufficient to give effective control until the end of the nursery cultivation, without any other chemical application. Only some mechanical or manual treatments are required.

An excellent result may be obtained also applying oxadiazon in autumn, following by a spring application of trifluralin + lenacil or dinitramin + lenacil.

To solve the particular problem of weed control in a strawberry nursery, effective results are obtained with applications of mixtures of lenacil + trifluralin or lenacil + dinitramin.   These treatments can be carried out at different periods but the best results are obtained with Fall or winter aplications. Another chemical application will be necessary in the spring and some soil cultivation treatments are also generally required.   A very interesting product is oxadiazon which is very effective even with one application only, followed by mechanical weeding.

The results obtained are very useful for field tests, thought further closely controlled experiments are to be continued.   Further tests which are now running are the complement of the tests we have been carrying out until now.

Session 4
Evaluation of herbicide tolerance

Chairman: H.Lawson

Improved methods of evaluating the tolerance of soft fruit crops to soil-acting herbicides

D.V.Clay
Weed Research Division, Long Ashton Research Station, Oxford, UK

Summary
Field evaluation of herbicide tolerance in the UK is often unreliable because results are affected by variation in rainfall and soil properties. A sand culture technique and field irrigation methods are more reliable. Both safe and damaging standard herbicide treatments must always be included.

Validity of the sand culture method has been confirmed by field experiments. The dose-response information obtained enables effects of overdosing to be predicted. The method also enables the tolerance of new crop cultivars to recommended soil-acting herbicides to be assessed.

In small-plot field experiments, irrigation was not always effective in inducing crop damage. Use of small plants, shallow planting and high herbicide doses can assist in obtaining positive results.

In interpreting likely safety of new treatments in the field, effects of crop age, growing system, root distribution, disease status and season of herbicide application must be considered. Differences in formulation and mixture with other herbicides can also affect response.

1. Introduction

Soil-acting herbicides have been widely used in soft fruit crops in the United Kingdom for the past 25 years. The safety of these herbicide treatments to crops has generally been established using small-plot field experiments. However because of variation in weather, particularly rainfall before and after herbicide application the phytotoxicity of treatments may vary greatly between experiments. This means experiments must be repeated in different conditions to obtain a reliable estimate of tolerance. In additon soil factors, particularly organic matter content, affect availability and downward movement of these herbicides so experiments must be carried out on different soil types to find generally safe herbicides and doses.

The effect of variation in weather on herbicide toxicity has been demonstrated in a series of experiments in which simazine was applied to young strawberries growing in a sandy loam soil at Begbroke Hill, Oxford (1, 2.) (Fig. 1). Although simazine is regarded as too damaging to be used in young strawberries in England, five out of the 13 applications caused no damage because the weather was dry after treatment.

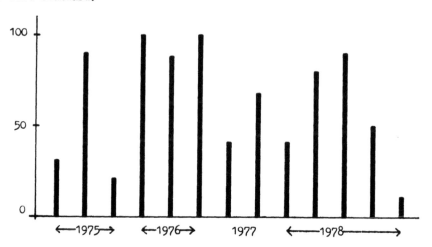

Crop vigour
(% of safe standard)

Fig. 1 The variable effect of simazine applied to young strawberries in
successive field experiments. Crop vigour recorded 1-2 months after
treatment.

These results have three consequences for field evaluation of new
herbicides. Firstly they emphasize the need for herbicide applications at
different dates/years to try and include the conditions likely to give
damage from a new herbicide. Secondly they illustrate the value of
including standard herbicide treatments in such trials: damage from a
normally safe treatment would indicate abnormal conditions, lack of damage
from a normally toxic treatment would indicate that conditions did not
provide a good test of tolerance. Having both standards helps
interpretation of the results. Thirdly, the variability due to weather
shows that more efficient evaluation systems are required which reduce the
need for time-consuming and expensive field testing of numerous new
herbicides. A sand culture method has been developed which gives rapid
and reliable information on relative tolerance thus reducing the number of
herbicides having to be tested in the field (3, 4, 5).

2. Sand culture evaluation method
     Crop plants are grown outdoors in pots containing non-adsorbent
silica sand and a series of doses of the test herbicides applied to the
sand surface. Drainage is restricted by standing pots in foil dishes
during the 4 to 6 weeks treatment period but waterlogging is avoided by
providing rain protection using a mobile transparent cover. Standard
herbicides of similar mode of action to the test herbicide are included.
From the results of the dose response series, doses causing particular
levels of effect (E D values) can be calculated. For tolerance studies an
E D 20 value is used (dose causing 20% growth inhibition). The relative
tolerance of different herbicides can be compared by calculating the
Tolerance Index, TI. (TI = ED value test herbicide/ED value standard
herbicide). E D values for a number of herbicides on strawberries are

shown in Table I (5). The slope of the dose/response curves can be compared by calculating a dose/response index, RI. (RI = ED 50/ED 20 value for each herbicide). Some examples are given in Table I.

Table I The response of strawberries in sand culture to applications of herbicides to the roots

Herbicide	Dose range tested (mg/pot)	Tolerance index*	Dose-response index*
Standard			
Simazine	0.14-26.0	1	2.0
Lenacil	1.20-97.2	4	1.3
Test			
Bentazone	0.32-20.5	10	7.0
Dimefuron	0.08-2.16	0.3	1.7
Ethofumesate	0.32-20.5	16	2.0
Metamitron	0.40-10.8	7	2.5
Oxadiazon	0.32-20.5	20	–
Oxyfluorfen	0.40-10.8	11	5.5
Pendimethalin	0.40-25.6	10	5.5
Propachlor	4.00-108	95	1.7

* Tolerance index, dose-response index – see text for explanation. Data from ref (5).

Table II Grouping of herbicides according to response to root and foliage applications in pot experiments and overall applications with commercial doses in the field

Herbicide	Pot experiments Root activity (sand)*	Foliage activity*	Field experiments
Dimefuron	●●●	o	●●●●
Simazine	●●	o	●●●
Lenacil	●	o	o
Metamitron	●	●	●●
Propachlor	o	●	o
Pendimethalin	o	●	●
Ethofumesate	o	●	●
Bentazone	o	●●●	●●●
Oxadiazon	o	●●●	●●●
Oxyfluorfen	o	●●●	●●●

* Key to response, o, no observable damage ----- ●●●● very severe damage
Data from ref. (5).

Photosynthesis inhibitor herbicides have low RIs indicating rapid increase in injury with dose, once the damage threshold has been reached whereas herbicide inhibiting root growth often give higher values. The response indices give valuable information on the likely effect of overdosing in field use. The relative tolerance of herbicides in sand culture has given a good indication of field performance, allowing for the different doses of herbicide used in field applications and the foliar-activity of some basically soil-acting herbicides (Table II) (5). As a result of this work and subsequent field testing a number of herbicides, including ethofumesate, pendimethalin and propachlor are now recommended on strawberries in the UK.

Although the results on relative tolerance in sand culture cannot be directly applied to field conditions the technique allows outstandingly safe herbicides to be selected quickly, similarly very toxic herbicides to be discarded. While the field response of herbicides of intermediate toxicity may be more difficult to predict the sand culture method used has the advantage over nutrient solution methods in that vertical distribution of herbicide is more like that occurring in the field (Table III), residues of most herbicides remaining near the sand surface. Information on the adsorption and soil mobility and persistence of the test herbicides can also be used to help decide if field testing is warranted.

Table III The distribution of herbicide in silica sand in 25 cm diam. pots, 1 day and 2 months after application of 4 mg a.i. to the surface

| | Residue recovered (ppmw)* | | | | | |
| | 1 day | | | 2 months | | |
Sampling depth (cm)	0-3	3-9	9-18	0-3	3-9	9-18
Lenacil	0.49	0.06	0.03	0.29	0.05	0.04
Propyzamide	1.24	0.23	0.03	0.80	0.13	0.10
Simazine	2.64	0.08	ND[+]	1.73	0.06	0.02
Terbacil	0.37	0.21	0.02	0.05	0.06	0.14
Trietazine	0.74	0.05	ND	0.22	0.04	ND

* Limit of detection 0.03 ppmw in all cases. Pots contained 8 kg sand (dry wt)
[+] ND = not detectable

Many soil-acting herbicides also have foliar-activity so this is normally tested in separate experiments on container-grown plants in which applied herbicide is kept off the soil during and after spraying (5). Many of the herbicides listed in Table II also showed foliar-activity which affected field response.

There are some types of herbicide activity which will not be accounted for in this type of pot trial. Phytotoxicity of herbicides from volatilization after application will not be allowed for nor will that of herbicides which can cause damage by splash of treated soil onto leaves near the ground e.g. oxadiazon and oxyfluorfen on blackcurrants (6). Certain chemicals only become herbicidal after application to the soil e.g. 2,4-DES and chlorthiamid. Such activity would probably be underestimated by the sand-culture method. Providing these limitations are allowed for the sand culture method can be a useful means of reliably obtaining information on crop tolerance independent of soil and weather variation.

3. Field evaluation using irrigation

An efficient field screening test is required to confirm the crop tolerance of promising herbicides. As with pot experiment techniques there is a need to induce damage from a recommended herbicide. Traditionally field evaluation has been done using small plots, normal and high doses of herbicide and a soil of low herbicide adsorptivity to increase chances of obtaining damage. Precipitation increases herbicide movement into soil, availability to crop roots and likelihood of crop damage. In England lack of activity of soil-acting herbicides due to dry weather after spraying is a frequent occurrence. Supplementing natural rainfall with irrigation is therefore commonly used to increase the reliability of field screening. This may be more useful than increasing herbicide doses since in dry conditions very high doses of adsorptive herbicides can be tolerated. In experiments with young strawberries plants have been damaged by doses of lenacil as low as 0.75 kg a.i./ha in wet conditions but unaffected by 22 kg/ha in a dry season. Activation of residual herbicides with irrigation requires uniform application and this is difficult to achieve due to inherent distribution differences with sprinkler equipment or effects of wind on deposition from oscillating spray lines. Overall irrigation may also not be appropriate in many experiments where planting or treatment dates vary from plot to plot. Small plot irrigation has been used to study herbicide toxicity to newly-planted strawberries with variable effectiveness (2). While in some experiments post-spraying irrigation increased the damage from lenacil and other herbicides (Table IV, V) in dry conditions heavy irrigation did not induce lenacil toxicity (Table VI) (2).

Table IV The effect of different soil moisture treatments on the response of strawberries to lenacil

Soil moisture treatments	Lenacil dose (kg a.i./ha)	Results as % untreated control		
		Plant vigour score, 2.5 MAT:	Crown no. 8.5 MAT	Dry wt leaves 7.5 MAT
Irrigated)		56	83	65
Natural)	0.7	107	104	89
Dry)		106	99	101
Irrigated)		5	32	11
Natural)	6.7	76	108	83
Dry)		92	90	114
Irrigated)	untreated	100 (7.8)*	100 (3.4)/	100 (118)+
Natural)	untreated	100 (7.5)	100 (3.3)	100 (105)
Dry)	untreated	100 (7.8)	100 (3.2)	100 (101)
S.E. + (soil moisture treatments at same dose)		5.3	8.2	6.8

: MAT, months after treatment
* 0-9 scale, 0 = dead 9 = healthy   / No./plant   + wt/plot, g

127

Table V The effect of soil moisture treatments on the response
of strawberries to different herbicides

Herbicide	Dose (kg a.i./ha)	Results as % untreated control			
		Plant vigour score 2 MAT[:]		Fresh wt leaves 2 MAT	
		Dry[+]	Irrigated	Dry	Irrigated
Chloroxuron	5.6	97	96	94	85
Metamitron	4.5	96	56	91	65
Simazine	1.5	100	48	100	65
Terbacil	0.25	89	85	95	76
Untreated		100	100	100	100
(Actual values)		(9.0*)	(8.9*)	(854 g/plot)	(945 g/plot)
S.E. $\pm$		2.5		11.3	

* 0-9 scale; 0 = dead, 9 = healthy   [:]MAT, months after treatment
[+] Plots protected from rain for 1 month after spraying

Table VI The effect of post-spraying irrigation on the response
of strawberries, cv Cambridge Favourite to lenacil (4 kg a.i./ha)
applied 1 month after planting

Herbicide treatment	Irrigation Depth (cm)	Results as % untreated, unirrigated control Leaf fresh weight (12 weeks post spraying)
Lenacil	0	100
"	1	105
"	2	108
"	4	113
"	8	115
Untreated	8	107
Untreated (actual value)	0	100 (64g/plant)

There are two particular problems with small plot irrigation in dry
conditions which reduce the chances of inducing damage. Firstly the
lateral seepage of water to surrounding dry soil probably reduces vertical
mass flow. Secondly to apply irrigation to many plots with a few machines
means application rates must be rapid (4 cm/hour in the work referred to
(2)). With herbicides of low water solubility such as lenacil and
simazine there is insufficient time for solution and downward movement of
significant quantities. Subsequent upward movement in solution in

128

response to surface water evaporation may also reduce final penetration depth. The conclusion from this and other work was that to obtain reliable results on relative tolerance uniform slow irrigation over a large area was needed.

Two ways of increasing the damage from soil-acting herbicides on strawberries were found which corroborate commercial experience. One was the use of smaller plants in tests, the other restricting the depth of planting of the roots to the top few centimetres of soil (2). Where damage from recommended herbicides is difficult to achieve such methods may be helpful.

4. Testing tolerance of new varieties to herbicides

Along with the need for information on new herbicide treatments, soft fruit growers need to know whether recommended herbicides are safe on new varieties. With some crops such as strawberries new varieties are frequently introduced and, in areas where growers rely on herbicides almost entirely for weed control, there is a pressing need for systematic

Table VII The effect of lenacil on strawberry varieties Cambridge Favourite (F), Senga Gigana (G) and Montrose (M) in sand culture and in the field

Lenacil dose (mg/pot)	Sand culture leaf dry wt (% untreated)			
	F	FS*	G	M
2.0	90	73	62	95
5.0	64	39	23	66
12.5	28	36	8	37
31.2	10	5	4	25
Untreated	100	100	100	100
(Actual value g/plant)	(10.1)	(7.0)	(8.4)	(4.7)
S.E. ± (treated v. untreated)		10.6		

Lenacil dose (kg a.i./ha)	Field			
	F	FS*	G	M
2.0	109	102	113	105
6.0	100	81	60	120
Untreated	100	100	100	100
(Actual value g/plot)	(75.8)	(69.0)	(63.8)	(57.2)
S.E. ± treated v. untreated		6.1		

* FS, small plants, C. Favourite

evaluation of tolerance. For soil-acting herbicides the sand-culture technique described earlier has been found to be a satisfactory method for ranking varieties in order of tolerance. With lenacil relative tolerance in sand has been similar to that in field tests. (Table VII) (7). As with testing new herbicides certain principles must be followed to gain maximum information from pot or field tests. Both a standard susceptible and a standard tolerant variety should be included in the list so that ranking of tolerance can be precise. Conditions need to be such that damage is caused to the tolerant variety so that the relative tolerance of the new variety can be identified. Unless the tolerant variety is damaged ranking of other undamaged varieties is uncertain. Damage can be obtained as with new herbicide testing, using a sand culture method or in field tests, by adjusting dose levels, and using irrigation.

Plant size of all varieties in a test must be typical of the commercial crop. Since small plants are more damaged this can have an overriding effect in varietal tolerance tests (7). Herbicides should also be applied at an appropriate time of year. For example, in experiments with ethofumesate in strawberries, there were bigger differences in varietal tolerance in winter, the recommended application time, than in spring. These differences appeared to be linked with the degree of winter dormancy of the different varieties (8). Varietal differences in growth rate and in root distribution in soil could also affect response to herbicides and need to be taken into account in assessing tolerance of new cultivars. In stone fruit greater damage from root application of simazine occurred on varieties coming into leaf earlier in spring (9).

5. Factors affecting results of tolerance tests

During work on developing methods of evaluating herbicide tolerance of fruit crops, a number of additional factors have been found to affect results and need to be allowed for in designing experiments, interpreting effects and predicting performance in commercial conditions. These include plant, cultural and herbicide factors.

Because of the change in crop tolerance with age and season, treatments that cause damage in young crops or from application in the growing season may still be safe to use on older plants or at other times of the year. Even treatments that may cause some damage may merit consideration if there is no alternative herbicide treatment for a particularly severe weed problem (5).

In strawberries the growing system can affect tolerance. Applications of propyzamide or simazine were more damaging to one year old crops grown as matted rows than as spaced plants (10) - further evidence of the effect of plant size.

In strawberries where overall spraying is unavoidable, application during the flower initiation period can have an effect on yield the next year (11).

Disease can affect the safety of soil-acting herbicides. In field experiments strawberries affected by Verticillium wilt have been shown to be damaged by recommended doses of simazine applied in autumn whereas healthy plants were unaffected (Clay, unpublished data)

Herbicide formulation affects response. In strawberries differences in effect have been shown between wettable powder and emulsifiable concentrate formulations of oxadiazon (12) and between wettable powder and suspension concentrate formulations of propachlor and simazine (13). Herbicide mixtures may also cause damage compared with individual components. In pot tests lenacil toxicity was caused on strawberries when applied in mixture with alloxydim sodium or fluazifop-butyl. (7)

130

The techniques outlined above demonstrate how tests of herbicide tolerance can be carried out more effectively. Further work is in progress to improve prediction of field tolerance on the basis of tests on container-grown plants.

REFERENCES
1. CLAY, D.V. (1978) The tolerance of young strawberry crops to a trietazine/simazine mixture. Proceedings 1978 British Crop Protection Conference - Weeds, 151-158.
2. CLAY, D.V. (1983) The effect of irrigation treatments on the phytotoxicity of soil-acting herbicides to strawberries. Aspects of Applied Biology 4 1983 Influence of environmental factors on herbicide performance and crop and weed biology, 403-411.
3. CLAY, D.V. and DAVISON, J.G. (1978) An evaluation of sand-culture techniques for studying the tolerance of fruit crops to soil-acting herbicides. Weed Research, 18, 139-147
4. CLAY, D.V. (1980) Indices and criteria for comparing the tolerance of strawberries to herbicides in dose-response experiments. Weed Research, 20, 91-96.
5. CLAY, D.V. (1980) The use of separate root and shoot tests in the screening of herbicides for strawberries. Weed Research, 20, 97-102.
6. CLAY, D.V. (1984) The safety and efficacy of new herbicide treatments for fruit crops. Aspects of Applied Biology 8, 1984 Weed control in fruit crops, 59-68.
7. CLAY, D.V. (1982) Evaluating the tolerance of fruit crops to herbicides: problems and progress. Proceedings 1982 British Crop Protection Conference - Weeds, 239-248.
8. CLAY, D.V. (1982) The tolerance of strawberry cultivars to ethofumesate alone or in mixture with lenacil or phenmedipham. Proceedings 1982 British Crop Protection Conference - Weeds, 291-298.
9. CLAY, D.V. (1984) Evaluation of the tolerance of cherry and plum trees to root application of herbicides using a sand-culture method. Aspects of Applied Biology 8, 1984 Weed control in fruit crops, 75-86.
10. CLAY, D.V. (1980) The influence of application date and growing system on the response of strawberries to propyzamide, simazine and trietazine + simazine. Proceedings 1980 British Crop Protection Conference - Weeds, 345-351.
11. CLAY, D.V. and ANDREWS, L. (1984) The tolerance of strawberries to clopyralid: effect of crop age, herbicide dose and application date. Aspects of Applied Biology 8, Weed control in fruit crops, 151-158.
12. CLAY, D.V. (1980) The effect of application timing and formulation on the tolerance of strawberries to oxadiazon. Proceedings 1980 British Crop Protection Conference - Weeds, 337-344.
13. CLAY, D.V. (1983) The relative toxicity of suspension concentrate and wettable powder formulations of herbicides to strawberries. Proceedings International Congress of Plant Protection, 1983, 578.

Assessment of herbicidal effect in relation to mode of action

J.C.Streibig
Department of Crop Husbandry and Plant Breeding, Royal Veterinary and Agricultural University, Copenhagen, Denmark

Summary
In principle there are two different methods of assessing the phytotoxicity of herbicides on the basis
of dose response curves.
    The first way of comparing the efficacy of two
herbicides is to measure the difference in test-plant
response at some preset dose levels.
    The second way of comparing herbicide dose-re-
sponses is to compare doses of each compound giving
similar response level. As a matter of fact, this
second method is in accord with the general theory of
selectivity used elsewhere in the biological sciences.
    The two methods of assessing the efficacy of
herbicides are discussed in relation to their mode of
action and the practical spraying situation in the
field.

1.1 Introduction

    Selective herbicides control uneconomic plants with-
out injuring economically important crop species. The selectivity can be assessed by comparing responses of different
plant species affected by one and the same herbicide or by
comparing the potencies of different herbicides on one and
the same plant species. Consequently, the measurement of
herbicidal selectivity is a relative measurement and is as
such a general concept for all biologically active substances.
    The purpose of this paper is to quantify herbicidal
effects by applying some general hypotheses about the
dose-response curves closely associated with the mode of
action of the herbicides in question.

1.2 The dose-response relationship

    When plotting plant production against the logarithm
of herbicide dose, the dose-response graph is often sigmoid extending from an upper limit, being similar to the
plant production unaffected by the herbicide, towards a
lower limit at large doses (Fig. 1). This sigmoid dose-response relationship appears to be general irrespective of
the mode of action of the herbicdes and the test plant
used.

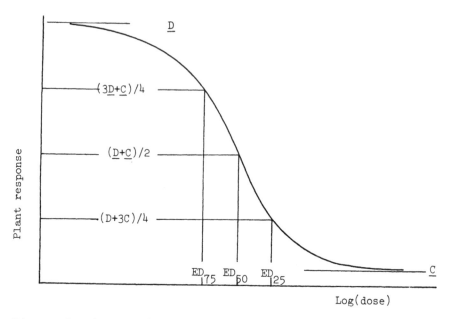

Figure 1. The meaning of the parameters in Eq. (1) describing the herbicide dose-response curve. The parameters $\underline{D}$ and $\underline{C}$ denote the upper and lower limit of the curve and are used to define the plant production which is equal to, for example, ED_{75}, ED_{50} and ED_{25}.

At subtoxic dose levels, however, most herbicides increase the plant production so that it exceeds that of the untreated control plants. This applies to almost all herbicides used so far in our laboratory, whether they are photosynthetic inhibitors, auxin type herbicides, inhibitors of meristematic growth etc.; but if the dose range is properly chosen to mask these subtoxic responses, the upper limit of the dose-response curves are always close to the untreated control (4).

In controlled experiments in greenhouse or/and in growth chamber it is fairly easy to avoid these subtoxic responses, but in field experiments, with a diverse weed flora, changing climatic conditions and different competitive conditions during the growing season, some tolerant weed species may grow vigorously and increase their production beyond the production of the very same weed species in the untreated control plots. The cause and effect relationship in those cases is far more difficult to interpret properly than under controlled environmental conditions, because two entirely different causes may be involved. Firstly, the tolerance of some weed species may be so that the doses tested really reflect a subtoxic effect. Secondly, the vigorous growth of apparently tolerant weed species might, however, be caused by improved competitive conditions, because their more susceptible fellow-weeds are killed or their growth reduced to such an extent that additional resources in terms of water, light, nutrients, space etc., become available for additional growth.

If the herbicide dose-response curve in Fig. 1 is accepted as an appropriate way of describing the action of a herbicide, in spite of the problems with subtoxic doses, we can succintly express the plant production (U) as a function of the herbicide dose (z) by

$$U=(D-C)/\{ 1+\exp(-2(a+b\cdot\log(z)))\} +C \qquad \text{Eq. 1}$$

This model describes the upper limit D at zero dose and the lower limit C at large doses (Fig. 1). The horizontal location is described by the parameter a, and b ($b<0$) denotes its slope half way between the upper and lower limits. By rearranging Eq. 1, we can obviously express the sigmoid curve by a straight line when knowing D and C;

$$Y=0.5\cdot\ln((D-U)/(U-C)) = -(a + b\cdot\log(z)) \qquad \text{Eq. 2.}$$

Y denotes a logit transformation of the original response (U). For example, the ED_{50} values in Fig. 1, which is defined as the dose required to reduce the plant production half way between the upper and lower limit of the curve, is equal to antilog($-a/b$) and this coincides with $Y=0$.

As the upper and lower limits of the dose-response curve as well as the ED_{50}-value are influenced by the condition under which the experiment has been conducted, it is certainly difficult to compare curves run under different environmental and/or experimental conditions, even though the test plant and the herbicide are the same.

Consequently, a dose-response curve, such as in Fig. 1, does not in itself explain anything, except that Paracelsus's 450 years old definition of the toxicity of substances is also valid for herbicides.

> "Was is das nit Gifft is: Alle Ding sind Gifft
> und nicht ohn Gifft.
> Allein die Dosis macht das
> ein Ding kein Gifft ist"
> (Paracelsus (1494-1541))

If we wish to assess the phytotoxicity of one and the same herbicide on differnt plant species or/and the effect of different herbicides on one and the same plant species, it is of course imperative to run the experiments under similar conditions or even better to run them simultaneously.

1.3 Assessment of herbicidal effect

In principle, there are two different ways of assessing the phytotoxicity of herbicides (Figs. 2 and 3).

Vertical assessment. The first way of assessing herbicidal effects concerns itself with comparison of plant responses at some preset dose levels. It is clearly shown in Fig. 2 that the dose levels greatly influence the differences of effects. If the tested doses are chosen in

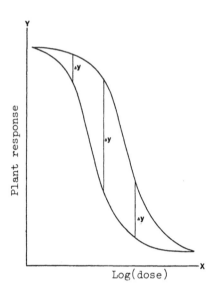

Figure 2. A vertical comparison of dose-response curves for two herbicides. Δy denotes differences of effect.

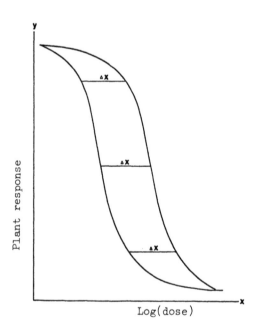

Figure 3. A horizontal comparison of dose-response curves for two herbicides. Δx denotes differences in log(dose).

136

such a way that the plant responses are close either to the upper or to the lower limits of the two curves, the differences of effects are less than if the doses are chosen in the middle. In particular, testing doses close to the upper or lower limits renders the difference in herbicidal activity invisible with a consequent lack of information about the intrinsic selectivity of the two herbicides. If the two dose-response curves in Fig. 2 are part of a factorial experiment, an analysis of variance will reveal significant interaction between the compounds, because the differences of effects are clearly dependent on the dose-levels used.

Horizontal assessment.The second way of assessing the efficacy of two herbicides is to compare the doses of each compounds giving similar effect (Fig. 3). In this case, we compare the horizontal location of the dose-response curves. As the doses are on a logarithmic scale in Fig. 3, this measurement of relative horizontal displacement expresses the ratio between the herbicide doses giving similar plant response. Mathematically, this can be expressed as the relative potency ($\underline{R}$) of the two compounds

$$\underline{R} = \underline{z}_s / \underline{z}_t.$$

Where $\underline{z}_s$ denotes the dose of a standard herbicide and $\underline{z}_t$ the dose of a test herbicide giving similar response.

In Fig. 3 it is noted that the relative potency, visualized by the horizontal lines, is more similar over a larger dose range than are the differences of effect in Fig. 2. This needs not always be the case, but in this particular instance, the two dose response curves are in fact parallel which gives a particular meaning to the relative potency as it is constant at any one response level considered.

Because of the asymptotic properties of Eq. 1 the plant responses can, as shown in Eq. 2, be represented by straight lines in Fig. 4 II and thus it is possible to talk about generalized parallelism, which is accomplished if the two curves are identical in all parameters expect for that causing the relative horizontal displacement (1). When using Eq. 2 one of the response curves can on a logit scale be expressed as

$$\underline{Y}_s = -(\underline{a} + \underline{b} \cdot \log(\underline{z}_s))$$

while the other curves can be described by

$$\underline{Y}_t = -(\underline{a} + \underline{b} \cdot \log(\underline{R} \cdot \underline{z}_t))$$

$\underline{Y}_s$ denotes the logit of plant responses for a standard herbicide and $\underline{Y}_t$ denotes the logit for the a herbicide; R being the relative potency which causes the horizontal displacement. In the present example $\underline{R}$ is independent of the response-level considered.

Fig. 4 shows two response curves for soil-applied technical grade ethofumesate (96% ai.) and a commercial product Nortron[r] (21% ai. of ethofumesate). The curves were

137

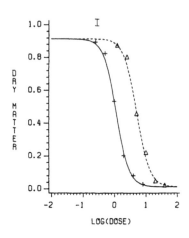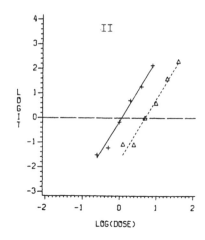

Figure 4. Dose-response curves of oat dry-matter on soil
-applied ethofumesate (+) and Nortron[r] (△); I shows original responses and II shows logit-transformed responses.
From (6).

considered to be parallel, because the compounds contained
the same active ingredient. The doses of ethofumesate were
based on mg ai./kg soil, whereas the doses of Nortron[r] were
based upon mg formulated product/kg soil. A parallel fit
in Fig. 4 appeared to be satisfactory from a statistically
point of view, and hence the relative potency was constant
at any one response-level considered. In this particular
experiment, the relative potency was 0.225 with a 95%
confidence interval of 0.021. From a practical point of
view this means that Nortron[r] contains 22.5% biologically
active ingredient of ethofumesate and is thus in accord
with the chemical declaration of 21%. A more detailed
description of this experiment is found elsewhere (6).

The parallel curves in Fig. 4 was of course brought
about by the different definition of the unit of doses;
for the technical grade ethofumesate it was defined as mg
active ai./kg soil whereas for Nortron[r] it was defined as
mg product/kg soil; but in both cases it was the very same
active ingredient that caused the effects.

Several experiments with different herbicides having
similar mode of action have shown that the hypothesis of
parallel dose-response curves is an appropriate way of
describing the curves. Table I shows some results from a
larger experiment that consisted of a total of eight dose
-response curves run simultaneouly. The response curves
for all four phenoxy acids had similar $\underline{D}$, $\underline{C}$ and $\underline{b}$ parameters whereas the $\underline{a}$ parameters were different because of
different horizontal location of the four curves. Obviously, the phenoxy acetic acids, MCPA and 2,4-D were more
potent than were the phenoxy propionic acids, mecoprop and
dichlorprop (table I). Within a new group of sulfonylurea
herbicides, we have also found parallel dose-response curves for root-absorbed chlorsulfuron, metsulfuronmethyl and

sulfometuron methyl when run simultaneously in greenhouse or growth chamber.

Table I. Summary of regression of _Sinapis alba_ dry-matter on the logarithm of phenoxy acid dose. Standard error is in parenthesis. Seedlings of S. alba were grown in growth chambers for fourteen days in a 650 ml aerated Hoagland nutrient solution with admixed herbicides. Each of the three randomised blocks consisted of six doses of each herbicide and two untreated controls.

$\underline{D}$ = 1.368 (0.217)		$\underline{C}$ = 0.014 (0.006)	
Herbicide	$\underline{Y}$ = -($\underline{a}$ + $\underline{b} \cdot \log(\underline{z})$)		ED_{50} mg ai./l.
MCPA	2.637 (0.232)	+ 2.081·log($\underline{z}$) (0.263)	0.054 (0.014)
2,4-D	2.589 (0.229)	+ 2.081·log($\underline{z}$)	0.057 (0.014)
Mecoprop	2.244 (0.203)	+ 2.081·log($\underline{z}$)	0.083 (0.022)
Dichlorprop	1.595 (0.177)	+ 2.081·log($\underline{z}$)	0.171 (0.045)

It is, however, important to stress that the hypothesis of parallel dose-response curves is a necessary, but not a sufficient condition for assuming similar action of the tested herbicides. For example, dose-response curves for TCA and chloridazon or TCA and methamitron have been found to be parallel, although the mode of action of TCA is different from that of chloridazon and metamitron (5).

1.4 Discussion
     The preceding chapter seems to have reduced the question of herbicidal effect and selectivity to a measurement of the relative potency of substances, which consist active ingredients having similar mode of action. This of generalization is probably difficult to evaluate under field conditions where climatic conditions vary through the growing season. Nevertheless, some out-doors experiments, which studied the effect of additives on herbicide performance, clearly showed that the effect of additives in the spraying solutions caused parallel displacements of dose-response curves for alloxydim-sodium and sethoxydim (7). Similarly, the effect of relative air humidity and temperature on herbicide activity in _Sinapis alba_ in growth chambers and the herbicidal efficacy on different developing stage of _Stellaria media_, _Matricaria inodora_ and _Myosotis arvensis_ ,grown in greenhouse were all properly described by assuming that the relative humidity, the temperature and the developing stage of the weeds only caused a parallel displacement of the dose-response curves.

REFERENCES

1. FINNEY, D. J. (1978). Statistical Method in Biological Assay. C. Griffin, London
2. KUDSK, P. (1985). Sprøjteteknik og additivers indflydelse på effekten af Avenge;(with English summary). 2 Danish Plant Protection Conference Weeds. 74-84.
3. KUDSK, P. (1985). "Faktorkorrigerede" doseringer - et alternativ til anbefalede doseringer?;(with English summary). 2 Danish Crop Protection Conference Weeds. 217-234.
4. STREIBIG, J. C. (1980). Models for curve-fitting herbicide dose response data. Acta Agriculturae Scandinavica, 30, 59-64.
5. STREIBIG, J. C. (1983). Fitting equations to herbicide bioassays: using the methods of parallel line assays for measuring the joint action of herbicide mixtures. Symp. Bioassay in Weed Science, Berichte aus dem Fachgebiet Herbologie der Universität Hohenheim, 24, 183-193.
6. STREIBIG, J. C. (1984). Measurement of phytotoxicity of commercial and unformulated soil-applied herbicides. Weed Research, 24, 327-331.
7. STREIBIG, J. C. and THONKE, K. E. (1985). The effect of a surfactant on alloxydim-sodium and sethoxydim potency. Symposium on Application and Biology, BCPC Monogram no. 28,147-154.

Session 5
Integrated pest management

Chairman: D.W.Robinson

Pest management as a basis foundation for integrated farming systems

A.El Titi
Landesanstalt für Pflanzenschutz, Stuttgart, FR Germany

Summary

Weed problems and the possibility for resolving them are
intimately associated with the whole producing system involved.
Changes of cultural techniques must be compatible with the
longterm goals of pest management, including those of weed
control. Changes in modern production systems which have taken place
during the last decades, particularly the increasing reliance
upon herbicides, have been adopted without sufficient knowledge
of the ecological effects concerned. Undesirable side-effects
such as inducing new pest problems by inhibiting natural
mortality factors, resistance to pesticides and environmental
pollution enforced a new orientation by farmers and consumers.
Integrated pest management is the most promising alternative
strategy. It emphasizes the exploitation of the natural control agents as a basic element in preventing pest population
to overstep the economic threshold. Increasing the antagonistic
agents and decreasing the susceptibility of the crop plants
are considered as framework for an integrated approach, in
which supervised control measures can be undertaken.

While evaluating the role of weeds, both noxious and beneficial
functions within the ecosystem involved are to be assessed. As
essential requisites (food and shelter), weeds can encourage the
survival of parasites and predators especially among terrestrial
arthropods and deflect initial attack of pest species. Control
measures against injurious weed species should be based on the
population dynamic data of the concerned weeds, as being
illustrated for Galium aparine. Mulching, green manuring and
soil tillage are discussed as cultural techniques supporting
Galium control. The economics of mechanical weed control compared with a herbicidal one is discussed;

1. Introduction

In recent years many farmers and scientists have become
aware that modern production systems in agriculture are

fragile, ecologically unstable. and showing increasing susceptibility to pest attack and weeds. And that is inspite
of the high progressive development in almost all agricultural disciplines. Disregarding ecological pest regulation
rules is suspected to be closely connected with the "modern
problems" (8, 9). Intensifying soil tillage for example,
especially ploughing, causes a repeated disturbance of
soil inhabiting organisms (4). Vast cropping of very few
varieties in a poor rotation decreased the vegetational
diversity. Intensive use of mineral fertilizers favoured
pests and diseases and encouraged nitrophilous weed species
to dominate. These measures have created pest and weed problems, forcing the application of more and more pesticides.
This intensive reliance on pesticides itself has induced new
difficulties, such as pest and weed resistance to some
pesticides (1).

2. The strategy to follow

How to face this new situation?

At least for the major pests, weeds and diseases, efficient
control techniques are available. This is not the main problem.
What we are looking for is a strategy, how to deal with the
available techniques, if the specific dangers are to be
avoided.

"Pest management" strategy was the next step in the development, making use of the specific data of life table of pest
species, before pest population increases. Specific antagonistic agents are managed to control a species of economic
importance, before causing economic losses. Difficulties and
limits became obvious, when more than one pest is involved.
The requirements for control of the one species conflicted
in many cases with those of the another one. Specific data
of the major species on the growing site are often not available

Different experiences being made in many countries have drawn
the practical "limits" of these new approaches. Especially,
within annual cropping systems, farmers deal with different
crops, with different pest problems and accordingly with
different natural antagonists, making it rather difficult to
fulfil the demands.

Integrated pest management in arable farming systems is a
new approach on this way. It builds a compromise for
population management of main pests in the concerned
cropping system (3). The most characteristic property is:

144

All techniques and measures of the production system should
be manipulated to increase the natural antagonistic
potential on the fields and on it's surroundings and to
decrease the crop susceptibilitiy to attack by pests, diseases and weeds. This is the framework of this approach,
regardless antagonistic agents against major or of whether or not
minor pest species are involved. Within this framework, supervised control including the application of pesticides can be carried
out to control single pest species. The ecological component of this approach is based on a long term continuation
or the integration of the ecological pest regulation components.

3. Main functions of weeds

Part of the instability of the agroecosystem can be linked
to the vegetational simplification, resulting from the
adoption of "pure monoculture", i.e. weedfree cultivation
without hedges or bordering shrubs (6). It is true that
vegetational diversity can be positive, negative of direct
or indirect effects on pest complex and accordingly on the
yields. Vegetation management, including that of weed
population should be based on the evaluation of weed
functions within the production system concerned, including
the crops to follow. By managing the population of some
weed species, herbivorous insect living on them can offer the
nutritional base for polyphagous predators or parasites
thus helping to suppress crop pests   (6). The weed functions
involved are mainly:

- Overwintering quarter:

   Many beneficial insects and mites find essential overwintering quarter in weed communities. Former infestation
   by phytophagous species facilitate the build up of antagonistic potential (7). Ground beetles, mainly carabids
   and staphylinids were found to be closely associated with
   weed communities.

- Alternative hosts:

   Van Emden (15) mentioned 442 references of weeds as
   alternative hosts for pest species. Most of the insects reported tended to feed on wild plants. For initial pest
   population or dispersing from the neighbourhood, it means
   deflecting the attack (13) or increase the natural enemies
   (2). Chenopodium album and Raphanus sp. germinating together with sugar beet reduced attack by a soil inhabiting collembolan pest - Onychiurus fimatus - resulting
   an increase of the establishmant rate (14).

Matricaria chamomilla in sugar beet fields exposed to the alates of _Aphis fabae_, attracted the aphids leaving sugar beet plants standing beside with very few or no aphids.

Matricaria attracted wireworm (_Elateridae_) larvae, reducing the pressure of attack on sugar beet seedlings.

The practical consequence of these results is to delay weed removal until completion of the susceptible youth stage. By the way, a significant yield depression can be expected, if weed removal is delayed more than 4 weeks after emergence (5, 11).

- Competition with crop plants:

Weeds are only harmful, if they decrease yields, or cause technical difficulties during or after harvest. Weed density at which level such losses are expected is called the economic threshold. In spite of having data about the economic threshold for different weed species in different crops, there are almost no practical knowledge on it's longterm effects.

How to get a population of a dominant weed species under levels of the economic threshold? It must be embedded in the whole farming system, while considering the ecological component mentioned.

4. Exploitation of the population dynamic data

"Cleavers" - _Galium aparine_ - is one of the most problematic weed species, not only in farm crops, but also in vineyards. Some population dynamic data (10) of this weed can help managing it's population in the long term aspect.

The rubiatic weed (Fam. Rubiaceae) increased on arable fields mainly because of

1. high N-dressing
2. disappearance of competitive weed species (selection)
3. application of unsatisfactory herbicides leaving _Galium_
4. no mechanical weed control, as manual hoeing.

Cleavers seeds - as being mentioned in different publications - do not survive more than 10 years in the average. Soil ploughing seems to reduce it's natural mortality. By turning soil up - side - down, seeds of deep soil layers come in a good position for germination. On the contrary, those of upper layer will be conserved until next ploughing.

The distribution of Galium seeds in soil profile one year after ploughing indicates an aggregation in the middle of soil profile. After a second ploughing the majority of the seeds are refound in the upper 5 cm layer (30 %). It means, these seeds are not affected by the weed control of the two previous years. The ability of seeds to germinate is almost the same in all layers. Field-emergence is located only between 1,5 - 3 cm. Seed germination and emergence out of deeper layers is extremely low.

To make use of this fact, soil should not be ploughed. It should be loosened with alternative implements - such as Chisel plough or 'broadshare cultivator'. If this is the case, seed embedded lower than 3 cm will be highly exposed to the mortality factors for a long time. Meanwhile, seed germination of the upper layer will be proportionally higher and the control measure would be more efficient.

The temperature optimum for Galium seeds to germinate is 7 - 13°C. That means the majority of seeds will germinate during autumn and spring. Timing weed control should consider this temperature demands, and is to be carried out after germination is completed.

Galium seeds need dark or semidark and humid conditions to germinate. Providing green manure cover would stimulate germination. An alternating mulching of green manure would help to reduce the seed bank in the soil.

Almost no data are available on the seed or seedlings mortality caused by pathogens or by soil animals. It is to be pointed out that the single control techniques mentioned cannot promise to provide a complete control, but in the system it might be able to reduce it's abundance to a tolerable level.

Establishing such an integrated management system in a commercial operating farm cropping cereals, sugar beets and vegetables shows some tendency to reduce Galium aparine after 6 years run. In the experiment mentioned, soil tillage, N-fertilizers supply, sowing technique and the direct control measures were manipulated to fulfil the demands of integrated pest management in arable farming systems. The result for cleavers is shown in table 1.

The key idea of this paper is to illustrate that crop protection largely depends on the consideration of the whole farming system. Changes in agricultural or horticultural practices are closely connected with changes on the status

147

of pests and weeds. If side effects of the modern production system are to be avoided or minimized, the integration of the ecological regulation on, under and beside the cropped plants is necessary. Manipulation of farming systems show the direction to go, but it needs a multidisciplinary cooperation. This is the hope of farmers and consumers.

REFERENCES

1. AMMON, H.U., E. BEURET (1984): Verbreitung Triazin-resistenter Unkräuter in der Schweiz und bisherige Bekämpfungserfahrungen. Z.Pfl.krankh.u.Pfl.schutz, Sonderheft X, 183-191
2. EL TITI, A. (1974): Überwanderung aphidophager Insekten von Erbsenfeldern zu benachbarten Feldern und ihre Bedeutung für den integrierten Pflanzenschutz. Z.Pfl.krankh.u. Pfl.schutz 81,287-295
3. EL TITI, A. (1984): Integrierter Pflanzenschutz, Modellvorhaben Ackerbau - Lautenbacher Hof. Hsg. Landesanstalt für Pflanzenschutz
4. EL TITI, A. (1984): Auswirkung der Bodenbearbeitungsart auf die edaphischen Raubmilben (Mesostigmata: Acarina). Pedobiologia 27,79-88
5. HACK, H. (1981): Vergleich der chemischen Unkrautbekämpfung in Zuckerrüben mit einer mechanischen Sauberhaltung zu unterschiedlichen Zeitabständen. Z.Pfl.krankh.u.Pfl.schutz, Sonderheft IV, 355-363
6. MAIER, C.T. (1981): Parasitoids emerging from puparia of Rhagoletis pomonella (Diptera: Tephritidae) infesting hawthorn and apple in Connecticut. Canad.Entomol. 113,867-870
7. MARTENS, B. (1983): Der Einfluß von Streifenanbau zwischen Hafer und Erbsen auf die Populationsdynamik der Getreideblattläuse (Homoptera: Aphididae) und ihrer Antagonisten. Diss. Heidelberg
8. NICKEL, J.L. (1973): Pest situations in changing agricultural systems - a review. Bull.Entomol.Soc.Am.19(3), 136-142
9. NORTON, G.A. (1984): Changing problems and opportunities for the adoption of integrated crop protection in cereals and associated crops. Vortr. EG-Konf.übr.integr.Pfl.schutz Brüssel, 9.-12.Okt.1984
10. RÖTTELE, M.A. (1980): Populationsdynamik des Klettenlabkrautes (Galium aparine L.). Diss. Hohenheim
11. SCOTT, R.K., S.J. WILCOCKSON, F.R. MOISEY (1979): The effects of time of weed removal on growth and yield of sugar beet. J.agric.Sci, Camb., 93,693-709

12. STEINER, H. (1980): Der integrierte Pflanzenschutz im Ackerbau in Baden-Württemberg. Gesunde Pflanzen 32,153-154
13. THEUNISSEN, J., Den Ouden, H. (1980): Effects of intercropping with Spergula arvensis on pests of Brussel sprouts. Ent.exp.appl. 27,260-268
14. ULBER, B. (1980): Untersuchungen zur Nahrungswahl von Onychiurus fimatus (Gisin) (Onychiuridae, Collembola), einem Aufgangsschädling der Zuckerrübe. Z.ang.Ent. 90, 333-346
15. VAN EMDEN, H.F. (1965): The role of uncultivated land in the biology of crop pests and beneficial insects. Sci. Hort. 17,121-136

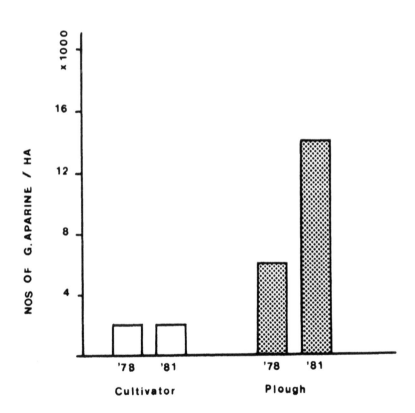

CHANGES WITHIN WEED COMMUNITIES IN RELATION TO SOIL TILLAGE
TECHNIQUES AFTER THREE YEARS RUN UNDER ARABLE FARMING CONDITION

	Mean No. of weeds per m^2			
	Cultivator		Plough	
	'79	'82	'79	'82
Thlaspi arvense L. (Field pennycress)	15	0	2.2	0.6
Cirsium arvense Scop. (Creeping thistle)	13.6	2.2	6.0	0.0
Polygonum convolvulus L. (Black bindweed)	7.4	11.0	0.6	0.4
Lamium purpureum L. (Red deadnettle)	1.8	12.0	0.0	2.8
Polygonum spp.	1.2	0.4	0.0	1.2
Alopecurus myosuroides H. (Black grass)	0.6	3.8	2.0	0.8
Senecio vulgaris L. (Common groundsel)	0.0	0.4	0.0	0.0
Veronica spp. (speedwell)	0.4	0.2	0.0	1.0
Stellaria media Vill. (Common Chickweed)	0.0	0.2	0.0	0.0
Convolvulus arvense L. (Field blindweed)	0.0	0.2	0.0	0.0
Matricaria chamomilla (wild chamomile)	0.0	0.0	0.2	0.0
Chenopodium album (White goosfoot)	0.0	0.0	0.2	0.4
Viola arvensis Murray (Wild pansy)	0.0	0.0	0.2	0.0
Anagallis arvensis L. (Scarlet pimpernel)	0.0	0.0	0.0	0.2
Galium aparine (Cleaver)	0.0	0.0	0.0	0.2
No. of spp.	7	9	7	9

Conclusions and recommendations

Summary

Where chemicals are used to control weeds, the long-term welfare of the consumer is of paramount importance. A greater standardisation of the law relating to the use of pesticides throughout the European Communities Countries is required.

Despite intensive herbicide usage, many problems with weeds remain in vine and soft fruits. In the last thirty years many new weed species have become prevalent as a result of intensive herbicide use and changing agronomic practice.

It was agreed that information from the ten Countries would be collated to determine the five most troublesome weed species in vine and soft fruits in the European Communities. It was also recommended that there should be a common research programme on the control of these weed species in the framework of the "Integrated Plant Protection" programme.

Alternative methods and means of controlling weeds other than by herbicides should be developed. The feasibility of integrated pest management (including weeds) also requires more attention at CEC level.

More studies are required on the management of weed populations. This may involve investigations on the use of living and dead mulches. The objective adopted in some countries of preventing all weeds from seeding in vines and fruit plantations requires critical economic evaluation. More information is required on the mode of action of some important herbicides so that improved levels of control can be achieved along with increased crop selectivity.

Considerable information exists on the yield reducing effects of weeds but little is known about which resource is limiting as a result of competition in specific situations. Studies on the survival strategies of problem weeds are required to provided a rational basis for their containment and management.

It was recommended that the Experts' Group on Vine and Soft Fruits (enlarged to include Top Fruits) should be re-convened in two years time to consider progress made in the work started or re-orientated as a result of the Dublin meeting. The Group suggested that other future meetings should consider "Weed control problems in modern methods of vegetable production" and "The problem of herbicide-resistant weeds".

In the light of the papers presented, the discussion, and the field visits, the following are the main conclusions and recommendations.

Good weed control is now essential for modern fruit growing and with the reduced resources for research now available in many countries, a determined effort will be required to achieve maximum efficiency in the research effort by pooling resources between countries of the Commission of the European Communities. It is evident that the problems concerning the control of weeds in vineyards and soft fruits are very similar and that the exchange of information between the two groups has been very useful.

There should be a more organised exchange of official weed control recommendations (label recommendations) between the ten Countries and also of information on the differences between cultivars in their susceptibility to herbicides.

Where chemicals are used to control weeds, the long-term welfare of the consumer is of paramount importance. More information on the residue levels resulting from herbicide treatments and on the possible side-effects of such residues is needed.

A greater standardisation of the law relating to the use of pesticides throughout the EC-Countries is required. In addition, the registration of herbicides should be the subject of a CEC directive.

Weed control strategies in vine and soft fruits

At the experts' meeting two different types of strategies emerged. In countries where wet wheather for prolonged periods could prevent weed control operations and where management skills with herbicides have reached a high level, a policy of weed control to prevent weeds from seeding has been adopted. In other countries weed management is the normal policy. Complete prevention of weeds from seeding in soft fruit crops can now be achieved in some areas and is a recommended objective in some EC-Countries (Great Britain and Ireland). Such an approach requires much attention to detail and is costly if assessed on a one-year basis only, but may be economic if evaluated over the life of the fruit plantation.

Because of the speed with which the weed flora can change, it is evident that a "weed-free" environment can only be maintained over a period of years with constant vigilance and continuing effort. The size of the weed seed population and longevity of weed seeds in the soil, the ability of the few surviving weeds to replenish that population and the ability of "opportunistic" species to exploit rapidly any open space indicate that weeds will never be completely suppressed. Where a virtual "weed-free" environment can be achieved, a number of significant economic benefits would accrue. All weed competition with the crop would be prevented and crop vigour and yield would increase. The spread of weed biotypes with acquired resistance to herbicides could not occur if weed seeding can be prevented. In addition, a small number of weeds in a fruit plantation may act as alternate hosts for nematodes and other crop pests. It seems likely that the aim of complete suppression of weeds, basically with herbicides (both overall and spot applications) but supplemented with some hand cultivation, will result in the use of a lower total amount of herbicides over say a 10-year period than a system aimed at achieving incomplete weed control only. Information is available from Scotland on the gradually reducing cost of weed control and amount of chemicals required following several years of successful herbicide treatments.

Investigations are required in a number of different localities to determine if it is possible to achieve "weed-free" crops with currently available methods. In addition, further information is required on the long-term effect of such treatments on soil properties including changes in organic matter levels.

On the other hand there are other situations where complete weed control is impracticable or undesirable. Here, more detailed studies will be required on the management of weed populations. Where weeds are being managed rather than controlled, information is required on the feasibility of maintaining populations of relatively non-competitive annual weeds in favour of perennial species. Work is required to determine if it is possible to prevent the development of difficult weeds including the role of living and dead mulches.

It is recommended that further investigation be undertaken to determine the validity for weed control of economic threshold levels in vine and soft fruits. This concept has been used effectively in the management of insects and nematode populations and also of weeds in extensively managed crops. However, less information is available on the suitability of this concept for weed control in fruits.

It is recognized that some weeds may have a useful role in harbouring beneficial parasites and predators of pests of fruit crops.

Where weed species and crops are grown together, the behaviour of weed species in the mixture is often different from the behaviour when the species are grown separately. Hence threshold levels in a particular crop will heve to take account of other weed species present also. Further many weeds will multiply by asexual as well as by sexual means and their development depends on a large number of factors such as crop density and soil fertility. Moreover, a few plants of a new weed species will not cause any economic damage in one season but if allowed to multiply could be very detrimental to subsequent crops. Some participants considered it ulikely that economic threshold levels as applied to insects and nematodes can be transferred directly to weeds in intensively grown crops such as soft fruits.

New weed problems

Despite intesive herbicide usage, many problems with weeds remain in vine and soft fruit crops.

A wide range of weed species has been controlled effectively, but many new weeds which were rare or uncommon in the 1950s and 1960s are now increasing in importance.

In addition, a number of common species have become resistant to herbicides to which they were previously susceptible. These include Epilobium ciliatum and Senecio vulgaris to triazines, and Poa annua to paraquat and triazines.

Changing practices in agriculture and horticulture has brought many new weeds to the fore. The increasing popularity of continuous minimally tilled winter cereal production has led to an increase in autumn-germinating weeds such as Galium aparine and Viola arvensis.

Many crops now occur as weeds in fruit plantations which were not a problem in pre-herbicide days. These include cereals, potato, oil seed rape, field beans, flower bulbs, pasture grasses, clover and fruit seedlings. These plants are frequently resistant to many of the herbicides used in soft fruits. Volunteer crops reduce the benefit of crop rotation by acting as a bridge for pests and diseases. Seedlings of the same species in a fruit plantation are especially difficult to control. More information is required on the development from seed of volunteer species and on their susceptibility to herbicides so that effective control can be developed.

It is recommended that joint action should be taken on the number of intractable weeds which now occur widely throughout the community and which are increasing in importance as a result of current herbicide and agronomic practice. Some of the important species which deserve special attention are Convolvulus arvensis, Trifolium repens, Epilobium ciliatum (American Willowherb) and soft fruit seedlings.

New herbicides

Despite the high cost of development, new herbicides are being introduced each year to meet the changing needs of modern agriculture. Herbicides are introduced by the chemical industry for large acreage crops only and very seldom is a herbicide introduced specifically for fruit growers. The work of the large chemical companies is directed only towards the large-volume markets and there is a continuing need for official research stations to examine the suitability for fruit growing of herbicides introduced for other crops and also the possible side effects resulting from the use of such herbicides.

Some perennial-weed problems (e.g. <u>Convolvulus</u> <u>arvensis</u>) are more likely to be solved by the introduction of completely new herbicides. There are opportunities for the introduction of new selective herbicides and new growth regulating chemicals for vine and fruit crops.

Although a high level of weed control has been achieved in many crops, our knowledge of the mode of action of many important herbicides is still incomplete. A better understanding of the action of herbicides under a range of environmental conditions and the specific sites in plants of their action or inhibitory effect, could do much to improve the level of control achieved and also increase crop selectivity.

Weed behaviour

Good weed management and effective weed control will require a much better understanding of weed response to changes in cultural methods as well as to the action of applied herbicides.

A weed population is not usually composed of a single weed species, but is often a higly dynamic plant community of perhaps 10-30 other species each attempting to suppress the crop and each other by different tactics and combinations of physiological and ecological strategies, such as methods of optimising resource capture and seed production.

Dearth of knowledge in many other aspects of weed ecology is considerable and there are many information gaps hindering the development of effective management practices.

Although we have considerable information on the yield reducing effect of weeds, little is known about which resource - water, light or nutrients - is limiting as a result of competition in specific situations. There is also little knowledge in perennial fruit crops of weed reproduction and spread. In particular, studies on the survival strategies of problem weeds are required urgently to provide a rational basis for their containment and management.

New and alternative approaches to weed management

Although chemical herbicides will remain an important means of controlling weeds in fruit crops, alternative methods are receiving attention.

The feasibility of integrated methods of pest management requires more attention at CEC level.

Further information is required on the effect of husbandry methods and on the use of biological control methods in vine and soft fruits, such as the use of mycoherbicides and parasitic plants.

There is a growing demand for fruits produced without the use of pesticides.

Because of the importance of weed control problems and the value of meetings of Experts for disseminating information, increasing knowledge and improving control strategies and thereby resulting in a reduction in the use of herbicides, and because of the success of this meeting of Experts in Dublin, participants ask the CEC to develop a programme of coordinated and common activities in Weed Control.

It was suggested that future meetings should consider "Weed control problems in modern methods of vegetable production" to be held in Germany, or "The problem of herbicide-resistant weeds" to be held in Great Britain.

It was also recommended that the participants in the Dublin meeting should re-convene in two years time to compare results of work on vine and soft fruits initiated or re-orientated as a result of this meeting. In addition, it was recommended that the scope of the work should be enlarged to include top fruit and that future participants should include representatives from the new EC Member Countries; the venue for this meeting should be one of the two new Member States.

Appendix – Results of a questionnaire on the worst weeds in vine and soft fruits

D.V.Clay

Weed Research Division, Long Ashton Research Station, Oxford, UK

Method

Representatives of each country were asked to send a list of the five weeds of economic importance which were the most difficult to control. Weeds were listed in order of importance (1 = most important).

Table 1 gives a list of weeds, abbreviated names and frequency of listing. Tables 2 and 3 present the detailed information country by country. In most cases correspondents gave information on strawberries separately from other fruit, in a few the soft fruit list includes strawberries. Some lists included more than five species. The number of countries from which returns were obtained were: for vine, 4; for soft fruit, 10; for strawberries 9.

Because of the differences in reporting, the numbers of occurrences given in Table 1 do not exactly represent the true number. Also information was not available from some countries. However, the list does provide an indication of the major problem weeds in Europe in these crops.

Conclusions

Some important points are clear.

In vines, perennial weeds are the major problem in all countries, in particular Convolvulus arvensis, followed by Cynodon dactylon and Cardaria draba.

In soft fruits, perennial weeds also are the major problem, particularly Convolvulus arvensis, Elymus repens and Cirsium spp.

In strawberries, annual weeds cause more ploblems, 17 species being listed: Poa annua is the worst problem, followed by Stellaria media and Senecio vulgaris.

The most frequent perennial weeds reported are Cirsium arvense, Elymus repens and Convolvulus arvensis.

Some respondents mention triazine-resistance of annual weeds as a special problem.

April 1986 is the date of reference of this questionnaire.

TABLE I – KEY TO SPECIES LISTED AND NUMBER OF OCCURRENCES IN LIST

Weed species	Abbreviations in Tables 2,3	Number of occurrences in lists					
		Vines		Soft fruit		Strawberries	
		A*	B*	A	B	A	B
Aegopodium podagraria	Aeg pod	–	–	1	2	–	–
Amaranthus lividus	Ama liv	1	1	–	–	–	–
Amaranthus retroflexus	Ama ret	1	2	–	–	–	1
Anagallis species	Ana spp	–	–	–	–	–	1
Artemisia species	Art spp	–	2	–	1	–	–
Barbarea vulgaris	Bar vul	–	–	–	1	–	–
Capsella bursa-pastoris	Cap bur	–	–	–	–	1	2
Cardaria draba (Lipidium draba)	Car dra	2	2	–	–	–	–
Chenopodium album	Che alb)	–	–	–	–	1	2
Chenopodium species	Che spp)	–	–	–	–	1	2
Cirsium arvense	Cir arv)	–	2	4	4	3	5
Cirsium species	Cir spp)	–	2	4	4	3	5
Convolvulvus arvensis	Con arv	4	4	4	5	2	4
Cynodon dactylon	Cyn dac	2	2	1	1	1	1
Cyperus rotundus	Cyp rot	–	1	–	–	1	1
Daucus carota	Dau car	–	1	–	–	–	–
Digitaria sanguinalis	Dig san	–	1	–	–	–	–
Echinochloa crus-galli	Ech cru	–	–	–	–	–	1
Elymus repens (Agropyron repens)	Ely rep	2	3	6	7	2	5
Equisetum arvense	Equ arv	1	1	1	2	–	1
Epilobium ciliatum	Epi cil)	–	–	1	3	–	1
Epilobium species	Epi spp)	–	–	1	3	–	1
Erigeron canadensis (Conyza canadensis)	Eri can	1	1	1	1	–	1
Galinsoga parviflora	Gal par	–	–	–	–	–	1
Galium aparine	Gal apa	–	1	1	1	1	1
Glechoma hederacea	Gle hed	–	–	–	1	–	–
Hypericum species	Hyp spp	–	–	–	1	–	–
Lamium purpureum	Lam pur	–	–	–	–	–	1
Poa annua	Poa ann	–	–	2	2	4	7
Polygonum aviculare	Pol avi)						
Polygonum persicaria	Pol per)	–	1	–	1	2	4
Polygonum spp (annual)	Pol spp)						
Portulaca oleracea	Por ole	–	–	–	–	1	1
Ranunculus repens	Ran rep)	–	1	3	3	–	–
Ranunculus species	Ran spp)	–	1	3	3	–	–
Rorippa sylvestris	Ror syl	–	–	–	–	1	2
Rumex crispus	Rum cri)	–	–	1	2	–	–
Rumex obtusifolius	Rum obt)	–	–	1	2	–	–
Senecio vulgaris	Sen vul	–	–	2	5	3	5
Solanum nigrum	Sol nig	–	–	–	–	–	1
Sorghum halapense	Sor hal	1	2	–	1	–	–
Stellaria media	Ste med	–	–	1	2	4	4
Taraxacum officinale	Tar off	–	–	–	–	–	1
Trifolium repens	Tri rep	–	–	1	2	–	–
Urtica dioica	Urt dio	–	–	1	2	–	–
Veronica persica	Ver per)	–	–	–	1	3	3
Veronica species	Ver spp)	–	–	–	1	3	3
Vicia species	Vic spp	–	1	–	–	1	1
Viola arvensis	Vio arv	–	–	–	–	1	2
Volunteer cereals	V**	–	–	–	–	1	1

*A = Number of times included in first three worst weeds;
*B = total no. of times listed

TABLE II - WEEDS OF <u>VINES</u> LISTED IN DECREASING ORDER OF IMPORTANCE
(1 = most important and diffcult weed to control)

Country	1	2	3	4	5	6	
Greece	Cyn dac	Con arv	Sor hal	Dau car	Cyp rot		
Italy	Con arv	Cyn dac	Ely rep	Cir spp	Sor hal	Art spp	
Switzerland Annual spp	Ama ret	Ama liv	Eri can	Dig san	Gal apa	Vic spp	
Perennial spp	Con arv	Car dra	Equ arv	Ely rep			
F.R. Germany	Con arv	Car dra	Ely rep	Cir arv	Ama ret	Ran spp	Pol spp

TABLE III - WEEDS OF SOFT FRUIT AND STRAWBERRIES LISTED IN DECREASING ORDER OF IMPORTANCE

(1 = most important and difficult weed to control)

CROP:	Soft fruit *					Strawberries *					
	1	2	3	4	5	1	2	3	4	5	6
Greece	-	-	-	-	-	Cyp rot	Pol spp	Cyn dac	Con arv	Ech cru	-
Italy	Cyn dac	Cir spp	Con arv	Art spp	Sor hal	-	-	-	-	-	
Switzerland Annual spp	Ste med					**v	Ste med	Ver per	Poa ann	Sen vul	
Perennial spp	Con arv	Ely rep	Ran rep	Aeg pod		Ely rep	Con arv	-	-	-	-
France	-	-	-	-	-	Por ole	Poa ann	Pol avi	Pol per	Con arv	Ana spp
F.R. Germany	Ely rep	Aeg pod	Ran spp	Gle hed	Con arv (Rum cri)	Sen vul	Ste med	Ver per	Vio arv	Ama ret	Ror syl (Cap bur)
Belgium	Ely rep	Poa ann	Sen vul	Bar vul	Pol per	-	-	-	-	-	-
Netherlands (A+)	Ely rep	Poa ann	Sen vul	Ste med	Ver spp	Ver spp	Ror syl	Vio arv	Lam pur	Poa ann	Che spp
Netherlands (B+)						Poa ann	Ste med	Che alb	Sol nig	Gal par	Ely rep
Denmark	Con arv	Cir spp	Ely rep	Sen vul	Hyp spp (Equ arv)	Sen vul	Poa ann	Cir spp	Tar off	Ely rep	-
England	Con arv	Ely rep	Eri can	Sen vul	Epi spp	Con vul	Cir ann	Gal spp	Ely off	Eri rep	-
Ireland	Gal apa	Tri rep	Ran rep	Ely rep	Vio arv	arv	-	apa	rep	can	-
N.Ireland	Rum obt	Urt dio	Cir arv	Tri rep	Epi cil	Ste med	Cap bur	Poa ann	Sen vul	Pol avi	-
Scotland	Cir arv	Equ arv	Epi spp	Sen vul	Urt dio	Cir arv	Ely rep	Sen vul	Epi spp	Poa ann	-

**v = Volunteer Cereals

A+ = annual cropping system; B+ = crops 2 years old or more

NB.: Elymus repens = Agropyron repens

* Where weeds are not included under strawberries, they are included in the soft fruit list.

List of participants

Belgium:

        BOXUS Ph.
        Station des Cultures Fruitières et Maraîchaires
        Chaussée de Charleroi, 234
        5800 Gembloux

        BULCKE R.
        Laboratorium voor Landbouwplantenteelt en Herbologie
        Fakulteit van de Landbouwwetenschappen
        Rijksuniversiteit
        Coupure Links, 653
        9000 Gent

Bundesrepublik Deutschland:

        EL TITI A.
        Landesanstalt fur Pflanzenschutz
        Reinburgstrasse, 107
        7000 Stuttgart

        SEIPP D.
        Versuchs- und Beratungsstation fur Obst- und Gemusebau
        Langfoerden
        Spredaer Strasse, 2
        2848 Vechta

Denmark:

        NOYE G.
        National Weed Research Institute
        Experiment Station in Weed Eradication
        Flakkebjerg
        4200 Slagelse

        STREIBIG J.C.
        Department of Crop Husbandry and Plant Breeding
        Royal Veterinary and Agricultural University
        Thorvaldsensvej, 40
        1871 Copenhagen

Great Britain:

CLAY D.V.
Long Ashton Research Station
Weed Research Division
Begbroke Hill - Yarnton
Oxford OX5 1PF

GREENFIELD A.
Agricultural Development and Advisory Service
Weed Research Organisation
Begbroke Hill-Yarnton
Oxford OX5 1PF

LAWSON H.M.
Scottish Crop Research Institute
Mylnefield - Invergowrie
Dundee DD2 5DA

Greece:

GIANNAPOLITIS C.N.
"BENAKI" Phytopathological Institute
Weed Department - Delta, 8
14561 Kiphissia (Athens)

PASPATIS E.A.
"BENAKI" Phytopathological Institute
Weed Department - Delta, 8
14561 Kiphissia (Athens)

Ireland:

BURKE J.
ACOT
Dublin Road
Enniscoethy (Co. Wexford)

RATH N.
Agricultural Institute
Soft Fruit Research Station
Clonroche (Co. Wexford)

ROBINSON D.W.
An Foras Taluntais
Kinsealy Research Centre
Malahide Road
Dublin

WATTERS B.
Department for Agriculture for Northern Ireland
Horticultural Centre
Loughgall (Co. Armagh)

Italy:

MAROCCHI G.
Osservatorio Regionale per le Malattie delle Piante
Via di Corticella, 135
40129 Bologna

MIRAVALLE R.
MOITAL
Via Melchiorre Gioia, 8
20124 Milano

SCIENZA A.
Istituto di Coltivazioni Arboree
Università degli Studi
Via Celoria, 2
20133 Milano

Netherlands:

NABER H.
Plant Protection Service
Centrum for Agribiologisch Onderzoek
Geertjesweg, 15
6706 Wageningen

VAN DER SCHEER H.A.Th.
Research Station for Fruit Growing
Brugstraat, 51
4475 Wilhelminadorp

Switzerland:

BEURET E.
Station Fédérale de Recherches Agronomiques de Changins
Route de Duillier
1260 Nyon

C.E.C.

CAVALLORO R.
Commission of the European Communities
"Integrated Plant Protection" Programme
Joint Research Centre
21020 Ispra (Italy)

Index of authors